Emerging Pattern in Plant Genomics and Marker Assisted Breeding

— *Volume -2* —

THE EDITORS

Bonipas Antony. J, holds a Bachelor's in Agricultural Sciences from Lovely Professional University, Punjab, and M.Sc. in Genetics and Plant Breeding from UAS, Dharwad. He is now pursuing Ph.D. at TNAU. He has received ICAR-JRF and CSIR-JRF fellowships and cleared CSIR-NET (2020) in Life Sciences and ASRB-NET (2021, 2023). Bonipas has contributed to academic publications and engages in international and national conferences.

Maruthi Prasad B. P. with a Bachelor's in Agricultural Sciences and M.Sc. in Genetics and Plant Breeding from UAS, Dharwad, is currently pursuing Ph.D. at UAS, Bangalore. He holds CSIR-JRF fellowships, passed CSIR-NET (2023) in Life Sciences, ASRB-NET (2021 & 2023), and UGC-NET (2022). Maruthi actively contributed to publications, organized national and international conferences, and engages in scholarly events.

Shobica Priya R, holds a Bachelor's and M.Sc. in Agricultural Sciences and Genetics and Plant Breeding, respectively, from TNAU. She is presently pursuing Ph.D. at TNAU. She has cleared the ASRB-NET (2021), and is a recipient of TNAU Merit Scholarship (2019) and the DST-INSPIRE fellowship (2021). Shobica has authored academic publications, book chapters and actively participates in international and national conferences.

Pallavi. M, holds a Bachelor's in Agricultural Sciences from UAS, Bangalore, and M.Sc. in Genetics and Plant Breeding from ANGRAU, Andhra Pradesh. She is now pursuing Ph.D. at UAS, Bangalore. She has received ICAR-JRF fellowship and CSIR-JRF fellowships and cleared CSIR-NET (2023) in Life Sciences and passed ASRB-NET (2021 & 2023). Pallavi has contributed to publications and engages in international and national conferences.

Sangamesh Nevani holds a Bachelor's in Agricultural Sciences from UAS, Dharwad and M.Sc. in Genetics and Plant Breeding from UAS, Bengaluru and received Doctoral degree from UAS, Dharwad. Presently working as a PVERA in PPV & FRA New Delhi. He has authored academic publications and participates in conferences.

Emerging Pattern in Plant Genomics and Marker Assisted Breeding

— *Volume -2* —

– Editors –

Bonipas Antony. J

Maruthi Prasad B. P.

Shobica Priya R

Pallavi. M

Sangamesh Nevani

2025

Daya Publishing House®
A Division of

Astral International Pvt. Ltd.
New Delhi – 110 002

© 2025 EDITORS

ISBN: 9789359199122

Publisher's Note:

Every possible effort has been made to ensure that the information contained in this book is accurate at the time of going to press, and the publisher and author cannot accept responsibility for any errors or omissions, however caused. No responsibility for loss or damage occasioned to any person acting, or refraining from action, as a result of the material in this publication can be accepted by the editor, the publisher or the author. The Publisher is not associated with any product or vendor mentioned in the book. The contents of this work are intended to further general scientific research, understanding and discussion only. Readers should consult with a specialist where appropriate.

Every effort has been made to trace the owners of copyright material used in this book, if any. The author and the publisher will be grateful for any omission brought to their notice for acknowledgment in the future editions of the book.

All Rights reserved under International Copyright Conventions. No part of this publication may be reproduced, stored in a retrieval system, or transmitted in any form or by any means, electronic, mechanical, photocopying, recording or otherwise without the prior written consent of the publisher and the copyright owner.

Published by : **Daya Publishing House®**
A Division of
Astral International Pvt. Ltd.
– ISO 9001:2015 Certified Company –
4736/23, Ansari Road, Darya Ganj
New Delhi-110 002
Ph. 011-43549197, 8130496929
E-mail: info@astralint.com
Website: www.astralint.com

Acknowledgement

The completion of "**Emerging Patterns in Plant Genomics and Marker-Assisted Breeding**" Volume II has been a collaborative effort that involved the dedication and contributions of many individuals. We extend our heartfelt thanks and appreciation to all those who have been part of this project.

Our sincere gratitude goes to the contributing authors who shared their expertise and insights in the various chapters of this book. Your commitment to advancing the field of plant genomics and marker-assisted breeding is evident in the quality of the content you've provided.

We would also like to acknowledge the reviewers who meticulously evaluated the chapters, offering valuable feedback and helping to maintain the high standards of this publication.

Our appreciation extends to our colleagues, mentors, and academic institutions for their support and encouragement throughout the process of compiling this volume. Your guidance and resources have been instrumental in making this book a reality.

Lastly, we express our deep thanks to our families and loved ones for their unwavering support and understanding during the long hours dedicated to this project.

It is our hope that *"Emerging Patterns in Plant Genomics and Marker-Assisted Breeding" Volume II* serves as a valuable resource for the scientific community and contributes to the continued Progress of plant breeding and genomics.

Thank you all for your contributions, dedication, and enthusiasm.

Sincerely

Bonipas Antony. J

Maruthi Prasad B. P.

Shobica Priya R

Pallavi. M

Sangamesh Nevani

Preface

Welcome to "**Emerging Patterns in Plant Genomics and Marker-Assisted Breeding**" Volume II. In this comprehensive volume, we delve into the exciting world of plant genomics and marker-assisted breeding, exploring the latest advancements, applications, and challenges in the field. The chapters in this book provide a wealth of knowledge for researchers, plant breeders, students, and anyone interested in the fascinating intersection of genomics and agriculture.

The journey through this book will take you on a tour of precision selection, multi-trait selection, genomic selection, and various innovative approaches to improving crop traits. We'll discuss the implications of marker-assisted selection in practical plant breeding, the utilization of wild relatives for crop improvement, and tools for exploiting genomic diversity. Our exploration extends to breeding for stress tolerance, disease resistance, pest resistance, nutritional quality, and much more.

As the editors of this volume, we hope that the insights shared by the contributing authors will not only inform but also inspire further research and advancements in plant breeding. The future of agriculture is intricately linked with our ability to harness the power of genomics, and this book is a testament to the progress we've made in this field.

We extend our gratitude to the dedicated authors who have contributed their knowledge and expertise to make this volume possible. We also thank our colleagues, mentors, and institutions for their support and encouragement throughout this project.

We hope that "Emerging Patterns in Plant Genomics and Marker-Assisted Breeding" Volume II serves as a valuable resource for your exploration of this dynamic and ever-evolving field.

Bonipas Antony. J

Maruthi Prasad B. P.

Shobica Priya R

Pallavi. M,

Sangamesh Nevani

Contents

Chapter–1

Precision Selection: Selection Theory of Marker-Assisted Breeding

S.Sneha[1], M.J Naveen Kumar[1], M. Sriram[1] and Devadharssini Obula Kannan*[2]*

**Corresponding author-: devadharssini@gmail.com*

1.1.Selection

The success of a plant breeding program highly depends on the selection of plants with the trait of interest. In classical selection, it generally takes 8-12 years to complete the breeding and selection cycle to release a single variety. Since the domestication starts, the selection is based on the phenotype only. Though the phenotypic selection is effective, it has many demerits. As the traits are controlled by genes and that variance is calculated on genetic variance, some sought-after traits are highly influenced by the environment. Those traits on selection, we may not get complete efficiency based on the phenotypic selection. This type of phenotype-based selection is generally called morphological marker-based selection. The precision of this selection is not merely cent percent.

To overcome these problems and to get the maximum efficiency of the traits through selection, certain markers have been developed. Those markers were called molecular markers. Marker-assisted selection, also known as marker-aided selection, is the term used to describe the selection process that is based on molecular markers. The precision of this selection is higher than the classical selection. The marker-assisted selection is the selection of plants of favorable characters on the analysis or based on the flagging marker with the trait of interest.

1.2. Marker-Assisted Selection

Molecular markers were developed, ushering in a new era of plant breeding selection. During the pre-molecular era, the selection did not harbor entire genetic diversity. However,the The unlimited genetic variety of the gene pool has been produced or made accessible thanks to molecular markers. The evolution of markers is about half a decade comprising from protein variants to DNA repeats. This evolution was mainly by reason of the high throughput of the marker and its cost efficiency. "The first true molecular markers to be introduced were allozymes, as the name itself defines it as a variant allele of a particular enzyme" (Schlötterer, C. 2004). This became very well known because of its neutral effect after the mutagenesis and this condition has become a constraint for allozymes. Though the mutation creates a neutral effect, in the period of natural evolution non-neutral effects have been created. This led to creating a hindrance for the allozymes to survey the variation.

This major constraint caused the creation of markers that are based on DNA. It surveys the variation in the DNA directly, instead of surveying the protein variation of an enzyme. This DNA marker is often termed as a "random DNA marker (RDM)" (Salgotra, R. K., & Stewart, C. N., 2020). The RDM contains non-functional DNA with the trait of interest. However, some RDM refers to functional markers, as these are present within the DNA of interest. (Fig.1.1)

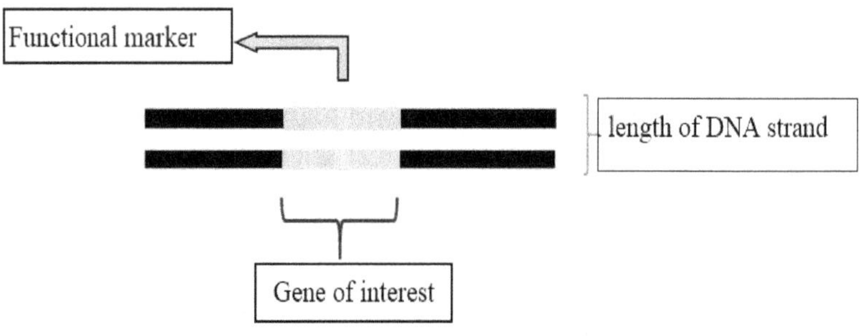

Fig. 1.1

"The evolution of markers was on the considerations of reliability, polymorphic level, technical procedure, requirement of template DNA, and cost" (Collard, B. C. Y., & Mackill, D. J., 2007).

❑ **Reliability-** The alignment of the markers with the target gene determines the reliability. Markers should be in less than 5cM, which provides a more reliable selection. In some cases, two markers may be present on either side of the gene of interest and this approach provides greater precision than using only one marker.

❏ **Polymorphism** – The marker should be able to distinguish the genetic differences in the set of individuals. Generally, "polymorphism is measured by heterozygosity" (Serrote, Caetano Miguel, et al., 2020)

❏ **Technical procedure** – A marker with simple procedures that saves time is mostly preferred. Markers with high throughput and the quickest methods are used. *e.g.* SNPs, KASP, etc.

❏ **Requirement of DNA–** some markers require a high amount of template DNA which leads to creating more copies causes more time consumption and high cost. The markers with low requirement of DNA are generally preferred by breeders.

❏ **Cost** – markers used in the selection procedure should be cost-efficient.

1.2.1. Markers

The functional marker identifies the optimal gene that controls the trait of interest. It is also referred to as a precision marker and the selection based on this marker is known as precision selection. In the functional marker selection/ precision selection the first part is the identification of the gene, affecting the phenotype of interest. This gene identification is usually done by approaches like expression profiling, map-based cloning, QTL mapping, and transposon mapping.

In recent times, this gene identification has become computerized which depends on the software under bio-informatics tools like,

Grial	Fgeneh/Fgenes
Genscan	Mzef
Procustes	Hmmgene
Geneid	Geneparser

Software Used for Gene Identification

Once the genes responsible for the trait of interest are identified, a study of variations in the allele of the respective gene characteristic is made. Then, the sequencing of the alleles of different genotypes is made for the comparative study of the polymorphism of a character. The most common polymorphism involves the variation at a single nucleotide and the other polymorphism involves larger variations involving long stretches of DNA such as microsatellite, VNTRs, etc. (Fig.1.2)

Single nucleotide polymorphism:

Maternal ....AGCTACGTTGAACCTGACGATTAG.......

Paternal ....AGCTACGTTGAGCCTGACGATTAG.......

Short tandem repeat sequence:

Maternal .....ACTGGACTGGACGACGACTCAT....

Paternal .....ACTGGACTGGACGACGACGACGACTCAT...

Fig: 1.2

Table: 1.1 (list of Common Molecular Markers)

AFLP	Amplified Fragment Length Polymorphism
RFLP	Restriction Fragment Length Polymorphism
RAPD	Random Amplified Polymorphic DNA
CAPS	Cleaved Amplified Polymorphic Sequences
EST	Expressed Sequence Tag
ISSR	Inter-Simple Sequence Repeat amplification
SCAR	Sequence Characterized Amplified Region
SSR	Simple Sequence Repeat
STS	Sequence Tagged Site
SNP	Single Nucleotide Polymorphism
REMAP	Retrotransposon-Microsatellite Amplified Polymorphism

1.2.2. Marker Characteristics

MARKER	UNIQUENESS
ALLOZYME	• Based on allelic variation of an enzyme. • It is quantified on the basis of bands formed during the protein electrophoresis. • Fast and inexpensive. • Limitations: biased highly on genomic sampling, low marker number, examination of significant sections of the genome, the sporadic discrepancy between tissues or ontogenetic stages, and challenges to standardize the experimental methods from laboratory to laboratory.
mt-DNA	• It is an extra-chromosomal mitochondrial genome. • As it follows maternal inheritance, used for comprehending the evolutionary history and constructing phylogenies. • Its drawback is that the analyses include hybridization, introgression, and incomplete lineage sorting.

RFLP	• Fragments of genomic DNA is created by restriction enzymes and the analysis of fragment is made.
	• Genetic variations in the fragments arise due to the nucleotide base substitutions, insertions, deletions, duplications, and inversions with the whole genome. Sequence information is not required.
	• Useful for finding out the specific gene location in a chromosome, genetic fingerprinting, profiling, or testing.
	• It is mostly used in the analysis of forensic science and several other fields.
RAPD	It is a PCR-based molecular marker.
	It does not require pre-sequencing and is a quicker and more effective way to screen for DNA sequence.
	A single, short oligonucleotide primer called RAPD, amplifies random sequences from a complicated DNA template by binding to multiple loci.
	Disadvantage: Only the presence or absence of a band based on a certain molecular weight allows for the detection of polymorphisms, with not much information on heterozygosity besides dominantly inherited, and also shows constraints with reproducibility of data.
AFLP	Rapidly generate hundreds of highly replicable markers from DNA.
	"Broad application in systematic, pathotyping, population genetics, DNA fingerprinting, and QTL mapping". (Al-Samarai, F. R., & Al-Kazaz, A. A. 2015)
Microsatellite	Variable number of tandem repeat and simple sequence length polymorphisms
	Ranges from one to six nucleotides in length.
	Used to study the linkage in families and linkage disequilibrium studies of populations, looks for the associations between the variation of allele and quantitative traits at neutral DNA markers
SNP	Developed by Lander (in 1996).
	Sequence polymorphism is caused by a single nucleotide mutation at a specific locus in the DNA sequence.
	It is a third-generation marker.
	Role: concerns with the population structure, genetic differentiation, origin, and evolution research.
	Disadvantage: Low-level information obtained in comparison with highly polymorphic microsatellites.

1.3 Marker Assisted Backcross Breeding

The transfer of one or few desired traits, (a transgene) from a donor parent into a favourable genetic background of a recurrent parent is termed to be backcross breeding. While introgression of a specific transgene, large amounts of undesirable donor genes in donor chromosomes remain, even after many generations of backcrosses. The agronomic performance of the plant under observation is negatively affected by these other donor chromosomes, hence they are deemed undesirable. This here is based on an assumption that DNA markers are able to predict the phenotype reliably. As the molecular markers are not affected by environment and have been more precise, marker assisted backcross breeding is more effective and quicker than conventional backcross breeding. This created a paradigm shift from the problems linked with conventional breeding by creating a variation in the selection criteria. This resulted in the selection of direct or indirect selection of genes that govern the desirable traits instead of selection of phenotypes.

Three stages are involved in Marker-assisted backcross breeding (MABB). They are "Foreground, Background and Recombinant selection". (K., SINGH, B. D. SINGH, 2016)

Foreground selection involves the selection of a sample or subset of plants that tends to carry a desirable allele. This desirable allele is obtained from the donor parent. Foreground selection is the genotyping of BC1F1 plants for Quantitative Trait Loci or a target gene using a linked marker. Here, a heterozygous state (one donor allele and one recurrent parent allele) of the target locus must be kept up until the final backcross is completed. Also a close linkage between marker locus and target locus is vital. The selected plants are selfed and the progeny plants are scrutinized to find the plants that are homozygous for the donor allele. Foreground selection, also referred to as positive selection was proposed by Tanksley. Foreground selection is used to selecting reproductive stage traits in seedling stages which in turn provides the best plants for backcross breeding and effective in resistance breeding for pests and diseases.

The next level of backcross breeding is the selection of a backcrossed progeny that holds the greatest proportion of the recurrent parent. Here, all of the genomic region is selected using the recipient parent's marker alleles and target locus is chosen on the basis of phenotype. This is achieved with the help of markers that are unlinked to the target gene/ QTL. Simply put otherwise, markers can be used to select against the donor genome. This selection is done to reduce the presence of unnecessary genes from the donor. This is referred to as complete line conversion.

The final stage is the selection of backcrossed progenies with the target gene. This is termed as recombinant selection which also is the recombination events between the linked markers and target locus. This is performed to reduce the introgression size of the donor chromosome segment that carries the target locus. Minimization of linkage drag can be achieved by flanking a target gene by utilizing a marker. This way, the recombinant selection in MABB is more effective than conventional backcross breeding as the donor segments can stay vast even after many backcrosses. It must be taken into consideration that recombinant selection is only possible for genes for which the map position is clearly charted. Recombinant selection reduces linkage drag, thereby posing effective for backcross breeding method.

1.4. Marker-Aided Selection's Far Arguments

Gene Pyramiding

Gene pyramiding is the introgression of two or more alleles of a gene of interest controlling a single trait from different plants into a single genotype. Mostly gene pyramiding is not only used to transfer genes of the resistance for the disease or insect but also to bring together many genes from the different loci that work well together. Servin et. al. (2004) proposed a scheme for accumulating genes from multiple parents into a single genotype. In normal gene pyramiding the selection is based on the phenotype. It would be difficult for the breeder to select the perfect genotype carrying the resistant gene. The marker-assisted selection made it easy to harbor the desired resistant gene. If we know the marker linked to the trait of interest, then the selection would be easy. It reduces the time taken for the selection of a trait and leads to the precision selection of the respective trait of interest.

There are trio schemes of backcrossing for gene pyramiding.

- ❏ "Separate backcross program
- ❏ Single backcross program - symmetrical mating
- ❏ Single backcross program - tandem mating" (K., Singh & B. D. Singh, 2016).

In a **separate backcross program**, genes from different donor parents (DP) are separately backcrossed with the recurrent parent (RP). Then the improved recurrent parent from each DP x RP crosses crossed together to accumulate all target genes in a single genotype. Then accumulate all target genes in a single genotype. On continuous selfing, coupled with the selection of complex hybrid gives the homozygous lines with all the target genes selected with the background of RP.

In a **single backcross program**, a single recurrent parent is crossed with the 'n' number of donor parents of desired traits. The F_1 from these crosses are crossed among them and the resultant progeny is backcrossed with the recurrent parent. This type of single backcross program is described as **symmetric mating**.

Another type of **single backcross program** is **tandem mating**. In this type, a single recurrent parent is crossed with a donor parent of the desired trait. The F_1 is crossed with another donor parent having a trait of interest. This procedure is repeated for future progenies with different donor parents. After introgressing all genes, the progeny is backcrossed with the recurrent parent to get a homozygous condition.

Gene pyramiding creates a possibility of losing the genes of interest due to recombination. Molecular marker overcomes this constraint and improves gene pyramiding of various traits like resistance to insect and disease, different morphological traits, etc. (Fig. 1.2)

Example: Pyramiding ASR-resistance genes (*Rpp1, Rpp2, Rpp3, Rpp4,* and *Rpp5)* through microsatellite (SSR) marker-assisted selection (MAS) in soybean and obtained plants containing from 2 to 4 genes pyramided per plant. For MAS, informative SSR markers in each cross. Joselaine Viganó (2018)

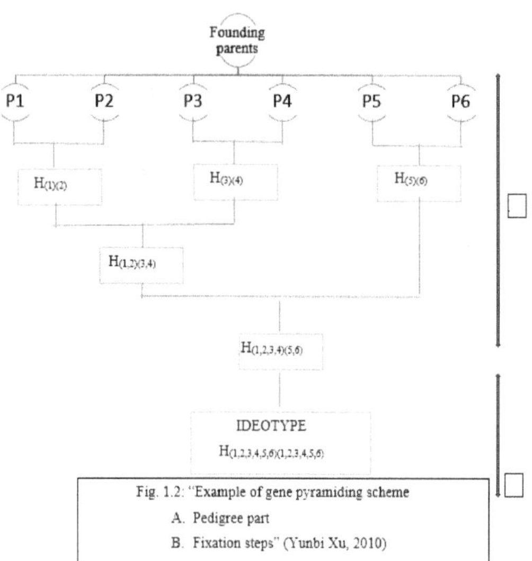

Fig. 1.2: "Example of gene pyramiding scheme

A. Pedigree part

B. Fixation steps" (Yunbi Xu, 2010)

1.5. Evaluation of Germplasm/Breeding Material

Markers are used for the purpose of easy or accurate selection. Germplasm materials used for plant breeding as a parental material are evaluated for the presence of respective desired traits. In those days, the phenotyping selection used to seek out the germplasms of desirable characters and it made a complex and time-consuming process. However, with the markers, an accurate selection of traits containing germplasm can be chosen. A high level of genetic diversity is required to exploit more heterosis. For example, in hybrid rice, the purity has been confirmed by SSR and STS markers. (Yashitola et al. 2002)

Selection is more important not only after crossing or hybridization but also for parental selection with a broad genetic base. The genetic diversity in the core-breeding materials should be increased, which would be useful to exploit the hybrid vigor. For this purpose, molecular markers serves as an important tool for assessing high genetic diversity plants in the germplasm. (Collard, Bertrand C.Y, and David J Mackill, 2007)

Year -Molecular marker -Crop/Species – Purpose Reference

1998 - RAPD- Mustard (*Brassica juncea)* - Genetic diversity (Rabbani et al., 1998)

2008 – RAPD- Lentil (*Lens culinaris*) -Genetic diversity (Sultana and Ghafoor, 2008)

2010 - RAPD- Cotton -Genetic diversity and germination pattern (Mumtaz et al., 2010)

For Agronomic Traits

Markers discriminate the different alleles of the targeted trait. So that the important agronomic traits can be selected on the basis of markers linked to the traits.

Table: 1.2. Markers for Selection of Some Agronomic Traits

CROP	TRAIT	GENE	CHROMOSOMAL NUMBER	MARKER SEQUENCE
WHEAT	Vernalization	*VRN-D1, VRN-H1, VRN-A1, VRN-B3*	5	F-CATAATGCCAAGCCGGTGAGTAC R-ATGTCTGCCAATTAGCTAGC
MAIZE	Flowering time	*Dwarf8*	1	F-ACACTATCACCGCTCTATTG R-ACTCTTTCCCTGACTTCATT
RICE	Semi-dwarf	*sd1*	1	F-CACGCACGGGTTCTTCCAGGTG R-AGGAGAATAGGAGATGGTTTACC
	Photoperiod-thermo-sensitive genic male sterility (PGMS and TGMS)	*pms3*	12	F-GAATGCCATCTAAACACT R-ATTTTACTCTTGATGGATGGTC
BARLEY	Photoperiod response	*Phd-H1*	7	F-CCTCTTCGCTATTACGCCAG R-GCCCTTCCCAACAGTTGCG

Note: Adapted from Kage, U., Kumar, A., Dhokane, D., Karre, S., &Kushalappa, A. C. (2015). Functional molecular markers for Crop Improvement. *Critical Reviews in Biotechnology, 36*(5), 917–930. https://doi.org/10.3109/07388551.2015.1062743

Table: 1.3. Markers for Selection of a few Quality Traits

CROP	GENE(S)	CHROMOSOMAL LOCATION	MARKER	TRAIT
MAIZE	Bm3	4	F-TTCAACAAGGCGTACGGGAT R-AGTGGTTCTTCATGCCCTCG	Forage quality of digestibility
	DGAT1-2	6	F-TGGCTCTGCAATCAGGAGAA R-TGAAGCAGCAAACAACGAGC	Oil content
SORGHUM	SbBADH-2	4	F-CGCAGTAGTGGAGTGGTTGT R-ACTGTGGCGGTTCTTGCATA	Fragrance
	SAI-1	4	F-GGATTCCACTTCCAGCCACA R-CGACGGGGTAGAAGTCGATG	Soluble acid invertase
MELON	gf	8	F-TCTGCAAAATGGTTGCTTTGAA R-AGGTGGATGTGGCACACAAA	Green fresh colour

Note: Adapted from Kage, U., Kumar, A., Dhokane, D., Karre, S., &Kushalappa, A. C. (2015). Functional molecular markers for Crop Improvement. *Critical Reviews in Biotechnology, 36*(5), 917–930. https://doi.org/10.3109/07388551.2015.1062743

References

Schlötterer, C. (2004). The evolution of molecular markers — just a matter of fashion? *Nature Reviews Genetics, 5*(1), 63–69. https://doi.org/10.1038/nrg1249 *Current protocols in human genetics.* (2011). Wiley.

Kage, U., Kumar, A., Dhokane, D., Karre, S., & Kushalappa, A. C. (2015). Functional molecular markers for Crop Improvement. *Critical Reviews in Biotechnology, 36*(5), 917–930. https://doi.org/10.3109/07388551.2015.1062743

Frisch, M., &Melchinger, A. E. (2005). Selection theory for marker-assisted backcrossing. *Genetics, 170*(2), 909–917. https://doi.org/10.1534/genetics.104.035451

Salgotra, R. K., & Stewart, C. N. (2020). Functional markers for Precision Plant Breeding. *International Journal of Molecular Sciences, 21*(13), 4792. https://doi.org/10.3390/ijms21134792

Al-Samarai, F. R., & Al-Kazaz, A. A. (2015). Molecular markers: An introduction and applications. *European Journal of Molecular Biotechnology, 9*(3), 118–130. https://doi.org/10.13187/ejmb.2015.9.118

Xu, Y., & Crouch, J. H. (2008). Marker□assisted selection in Plant Breeding: From Publications to practice. *Crop Science, 48*(2), 391–407. https://doi.org/10.2135/cropsci2007.04.0191

Hromadová, Z., Gálová, Z., Mikolášová, L., Balážová, Ž., Vivodík, M., &Chňapek, M. (2023). Efficiency of RAPD and Scot markers in the genetic diversity assessment of the Common Bean. *Plants, 12*(15), 2763. https://doi.org/10.3390/plants12152763

Collard, B. C. Y., & Mackill, D. J. (2007). Marker-assisted selection: An Approach for Precision Plant Breeding in the twenty-first century. *Philosophical Transactions of the Royal Society B: Biological Sciences, 363*(1491), 557–572. https://doi.org/10.1098/rstb.2007.2170

Qi, L., & Ma, G. (2019). Marker-assisted gene pyramiding and the reliability of using SNP markers located in the recombination suppressed regions of sunflower (*Helianthus annuus* L.). *Genes, 11*(1), 10. https://doi.org/10.3390/genes11010010

Tanksley, Steven D. "Molecular markers in plant breeding." *Plant Molecular Biology Reporter,* vol. 1, no. 1, 1983, pp. 3–8, https://doi.org/10.1007/bf02680255.

Ribaut, J.-M., et al. "Simulation experiments on efficiencies of gene introgression by backcrossing." *Crop Science,* vol. 42, no. 2, 2002, p. 557, https://doi.org/10.2135/cropsci2002.0557.

Hasan, Muhammad Mahmudul, et al. "Marker-assisted backcrossing: A useful method for Rice Improvement." *Biotechnology & Biotechnological Equipment*, vol. 29, no. 2, 2015, pp. 237–254, https://doi.org/10.1080/131 02818.2014.995920.

Serrote, Caetano Miguel, et al. "Determining the polymorphism information content of a molecular marker." *Gene*, vol. 726, 2020, p. 144175, https:// doi.org/10.1016/j.gene.2019.144175.

K., SINGH, B. D. SINGH, A. *Marker-Assisted Plant Breeding: Principles and Practices.* SPRINGER, INDIA, PRIVATE, 2016.

Viganó, J., Braccini, A. L., Schuster, I., & Menezes, V. M. (2018). <B>microsatellite molecular marker-assisted gene pyramiding for resistance to Asian soybean rust (ASR). *Acta Scientiarum. Agronomy*, 40(1), 39619. https://doi. org/10.4025/actasciagron.v40i1.39619

Collard, Bertrand C.Y, and David J Mackill. "Marker-assisted selection: An Approach for Precision Plant Breeding in the twenty-first century." *Philosophical Transactions of the Royal Society B: Biological Sciences*, vol. 363, no. 1491, 2007, pp. 557–572, https://doi.org/10.1098/rstb.2007.2170.

Ahmad, F., A. Akram, K. Farman, T. Abbas, A. Bibi, S. Khalid and M. Waseem. 2017. Molecular markers and marker-assisted plant breeding: current status and their applications in agricultural development. *Journal of Environmental and Agricultural Sciences.* 11: 35-50.

Chapter-2

Implication of Marker Assisted Selection: Practical Approach

T. Nivethitha[1*] and R. Nivedha[1]

Department of Genetics and Plant Breeding, Tamil Nadu Agricultural University, Coimbatore - 641 003
**Corresponding author: nivethitha120897@gmail.com*

2.1. Introduction

The current scenario of expanding global population is projected to reach nine billion by 2050, placing a growing demand on food production. Furthermore, challenges such as biotic and abiotic stresses, diminishing arable land and shifting climate patterns have compounded the pressures on crop production systems. Consequently, there is a pressing need for plant breeders to accelerate the breeding process in order to introduce improved varieties that offer enhanced productivity with resistance to biotic and abiotic stresses. This ultimately leads to the adoption of integrated approaches that combine conventional and molecular breeding methods to shorten the breeding cycle for developing new cultivars with desired traits.

The selection of superior genotypes is the preliminary step in a conventional plant breeding programme. The selection process is largely focused on phenotype. However, phenotype is the outcome of the genotype, environment and its interaction. Therefore, direct phenotypic selection is less efficient for quantitative traits. Herein lies the function of marker assisted selection (MAS), which allows for indirect selection of trait based on the molecular markers that are linked to it. Molecular breeding encompasses the utilization of molecular marker data to enhance the effectiveness of various breeding activities, including the planning and execution of breeding programs, as well as increasing the selection efficiency. This chapter covers the molecular markers available, various forms of MAS and provides a concise overview of the accomplishments achieved through MAS.

2.2. Molecular Markers

A specific characteristic feature of the individual that is heritable is termed as trait or character. Whereas, a marker is any landmark in the genetic makeup which segregates according to mendelian law together with the trait of interest. These markers are classified into four main groups as follows,

2.2.1. Morphological Markers

These are naked eye polymorphisms such as growth habit, colour and shape of flowers, fruits and seeds and other agronomic characters [1]. Though scoring of theses markers is rapid and simple, number of polymorphic markers in the crop species is less and it is highly influenced by environment [2].

2.2.2. Biochemical Markers

These are protein-based markers (isozymes) that can be easily detected as electrophoretic variants due to difference in the net charge or confirmation. Isozymes are the variants of an enzymes encoded by different genes. These are co-dominant markers frequently employed in population genetics. However, they are limited in number, show less polymorphism, and predominantly affected by plant tissue being used, growth stage, and method of their extraction. However, their utility is constrained due to lesser number of polymorphic markers, and influenced by factors such as the type of plant tissue used, developmental stage, and environment [3].

2.2.3. Cytological Markers

These markers are related to variations in chromosomes *viz.*, number, size, shape, position and banding pattern. Karyotyping and cytogenetic studies reveals the differences in the morphology and distribution pattern of heterochromatin and euchromatin [1].

2.2.4. Molecular Markers / DNA Markers

These markers are related to variations in the DNA sequences of the individuals. The polymorphism between the individuals is due to insertion, deletion, duplication, translocation and point mutations in nucleotide sequence. Molecular markers don't always directly target specific genes; rather, they are inherited as a sort of 'flag' that travels together with the gene of interest as traits passes from one generation to the next. They are also known as gene tags. An ideal marker should be, codominant, highly polymorphic, reproducible, inexpensive, distributed throughout the genome, high throughput and no environmental influence. Over the decades, in agricultural crops, many DNA markers has been successfully developed and utilized in plant breeding activities. They are classified into three groups based on their method of detection as follows [1].

- **Hybridization based markers:** Southern hybridisation, Restriction Fragment Length Polymorphism (RFLP), Diversity Array Technology (DArT)

- **PCR- based markers**: Randomly amplified polymorphic DNA (RAPD), Amplified fragment length polymorphism (AFLP), simple sequence repeats (SSRs), cleaved amplified polymorphic sequences (CAPS), sequence characterized amplified region (SCAR) and inter simple sequence repeat (ISSR) [4].

- **Sequence based markers:** Identification of order of base sequence in the DNA fragment is called DNA sequencing. In the wake of the enormous genomic advances, modern sequencing techniques like genotyping by sequencing, next- generation and third generation sequencing helps to detect large array of polymorphism at nucleotide level.

2.3. Marker Assisted Selection (MAS)

Since relying solely on phenotype-based selection isn't always a dependable indicator of the underlying genotype, plant breeders have long aimed to achieve indirect selection of traits of interest. This is where Marker-Assisted Selection (MAS) comes into play. MAS involves the selection of the desired allele of a gene or quantitative trait locus (QTL) based on molecular markers linked to the allele producing the desirable phenotype. Use of molecular markers in selection not only enhances precision but also significantly improves efficiency. The flowchart below illustrates the stages involved in the development and validation of markers and their subsequent utilization in MAS [5].

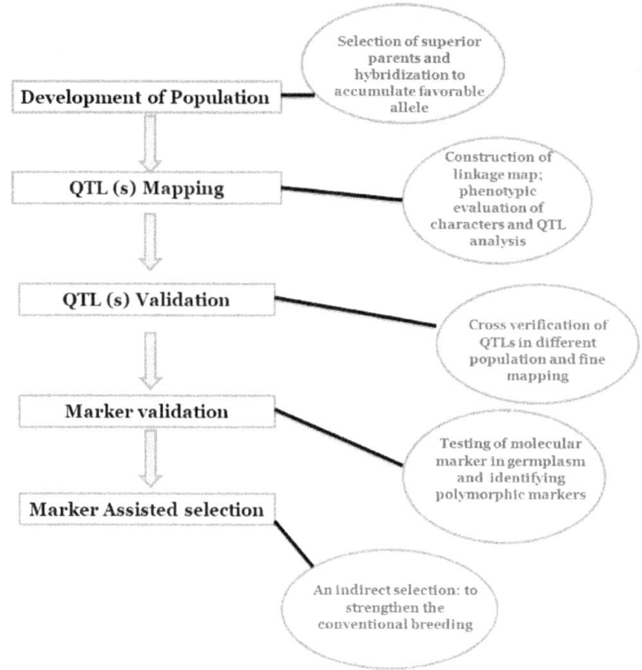

2.3.1.Forms of MAS

Various molecular techniques fall under the umbrella of MAS, including marker-assisted backcrossing (MABC), marker-assisted recurrent selection (MARS), gene pyramiding and genomic selection (GS) [1]. These techniques have been exploited in plant breeding for genetic material characterization and to exercise selection in the early generations, thus expediting the breeding cycle with enhanced precision.

2.3.1.1. Marker Assisted Backcrossing (MABC)

It is a technique that combines traditional backcrossing with molecular markers [6]. In MABC, a plant with a desired trait (donor parent) is crossed with a recipient parent, which usually has other favourable traits but lacks the specific trait of interest. After the initial cross, molecular markers linked to the target trait are used to identify and select true hybrids that have inherited the desired trait from the donor parent. The hybrids are then repeatedly backcrossed with the recipient parent for three to four generations to transfer the desired trait while retaining the valuable characteristics of the recipient parent. Based on the foreground selection for the desirable trait and background selection to increase the recipient parent genome recovery, individuals are selected and forwarded for next generation of backcrossing. MABC accelerates the breeding process by allowing breeders to focus on the specific trait of interest while maintaining the genetic makeup of the recipient parent. It enhances the efficiency and precision of trait introgression in plant breeding programs.

The basic steps in MABC programme are as follows,

- ❑ **Foreground selection**: using tightly linked marker for target locus selection *i.e.*, the indirect selection for the trait of interest to be transferred from the donor parent [7].

- ❑ **Recombinant selection**: using flanking markers of the target locus for selection to minimize the linkage drag.

- ❑ **Background selection**: using unlinked markers distributed throughout the genome to accelerate the recurrent parent genome recovery [7].

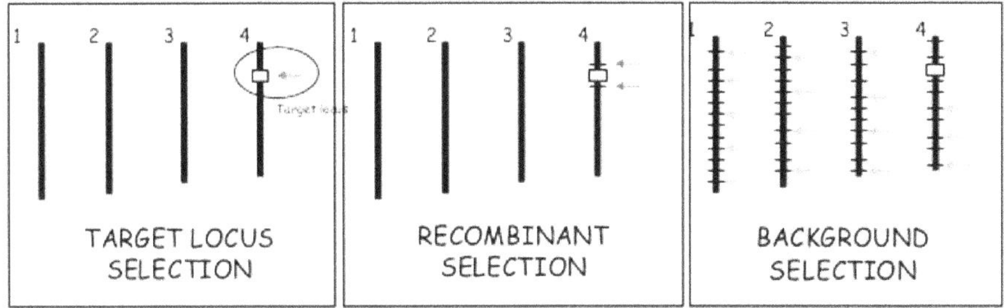

Fig.1: Foreground and Background Selection in MABC [8]

MABC is the frequently employed technique various crops *viz.*, rice, maize, pearl millet, sorghum *etc.*, The successful application of MABC has resulted in the release of numerous improved varieties for cultivation, and specific examples of these will be elaborated upon in the later section of this chapter.

2.3.1.2. Marker Assisted Recurrent Selection (MARS)

When much of the variation is controlled by minor QTLs, MABC has limited applicability because estimates of QTL effects are inconsistent and pyramiding becomes increasingly difficult as the number of QTLs increases. MARS is an efficient technique that increase the frequency of favorable marker alleles in the population by continuous crossing and selection process. The F_2 or F_2 derived progenies of a particular cross are phenotyped for target trait and genotyping is carried out using large number of markers distributed throughout the genome. The multiple regression analysis based on phenotype and genotype data results in the formulation of selection index. The individuals are selected based on selection index and are recombined and the procedure is repeated for number of cycles. It results in the accumulation of favorable alleles in the population and accelerates the selection cycle. One shortcoming of MARS is caused by the inconsistency of QTL effects as the genetic background changes during subsequent cycles of selection. The usage of this technique is limited since there is no advantage of MARS over phenotypic selection for a multitrait performance index, probably due to the general high heritability of traits and the limited phenotypic variance accounted by the QTL [9].

2.3.1.3. Gene Pyramiding

In general, gene Pyramiding refers to introgression on two or more genes governing a single trait from different donor parents to single recurrent parent.

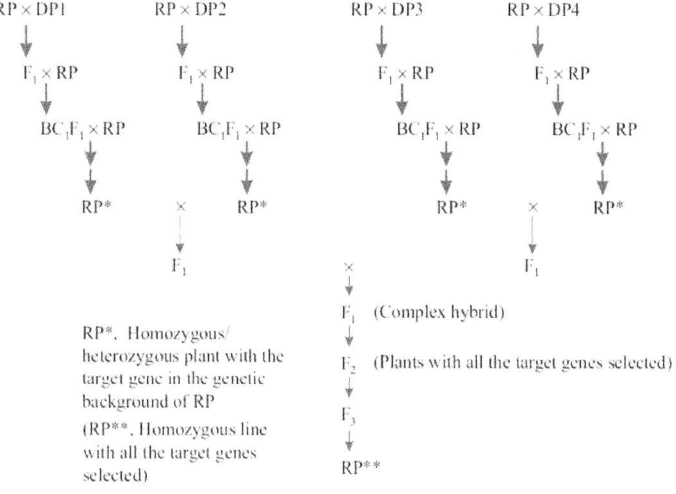

Fig.2: Schematic Representation of Seprate Backross Programme

In contrast, multitrait introgression refers to introgression on two or more genes governing different traits from different donor parents to single recurrent parent. When the objective is to introgress genes from different donor parents, three strategies are commonly employed: separate backcross programs (Fig.2.), symmetrical mating, and tandem mating (Fig.3) [2].

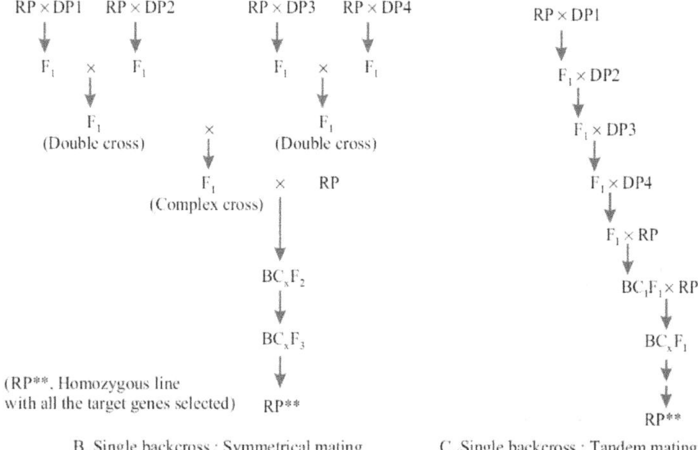

Fig.3: Schemmatic representation of symmetrical mating and Tandem mating

Among these approaches, the separate backcross program stands out as the most effective strategy, offering increased efficiency.

2.3.1.4. Genomic Selection

This is a specialized variant of MAS that relies on Genomic Estimated Breeding Values (GEBV) as the foundation for selection. GEBV, which is calculated using a Genomic Selection (GS) model, represents the cumulative effects associated with marker alleles of the genotyped individual [9]. GS involves two distinct populations, namely the training population and the breeding population, which are interrelated. The key steps in a GS program are as follows (Fig.4) [2]:

- ❐ **Training Population**: This initial population undergoes phenotypic evaluation, followed by genotyping using large number of markers that cover the entire genome.

- ❐ **GS Model Training**: The GS model is developed using both the phenotype and genotype data obtained from the training population. This process involves estimating the effects associated with all marker alleles.

- ❐ **Breeding Population**: In this population, individuals are genotyped using the same markers employed in the training population, but without any accompanying phenotypic evaluation.

- ❐ **GEBV Calculation**: Using the marker data from the breeding population and the effects associated with marker alleles obtained from the GS modeling, GEBV for each individual are computed. Selection of individuals is then based on these GEBV scores.

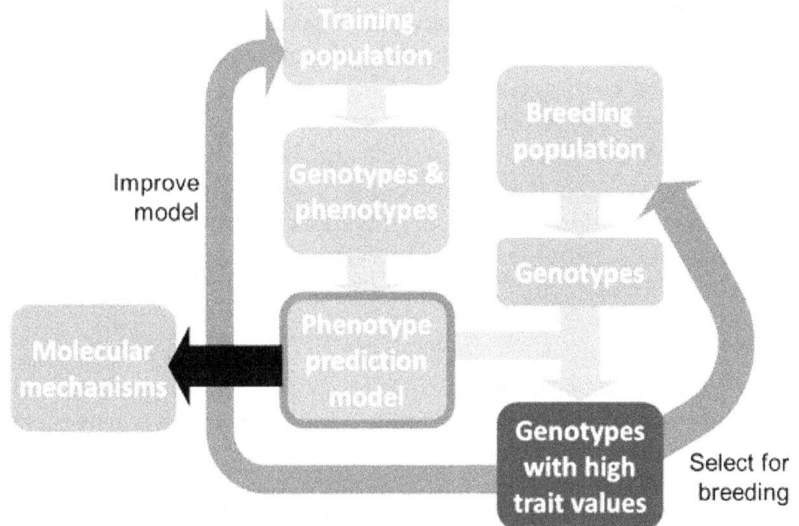

Fig.4: Genomic Selection Scheme

2.4. Achievements Through MAS

2.4.1. Private Sector

In recent decades, there has been a significant influx of private-sector investments in MAS, which is evident from the increasing number of patent applications related to molecular markers [10]. While a handful of MAS-varieties in field crops have been traced, such as drought-tolerant maize varieties like Agrisure Artesian and Optimum AQUAmax, low linoleic acid soybeans like Vistive, and stripe rust-resistant wheat cultivars such as Expresso, Blanca Grande 515, Summit 515, and New Dirkwin, it is important to recognize that many more potentially successful MAS applications likely exist [11]. However, these remain undisclosed due to the confidential nature of operations within private companies worldwide.

2.4.2. Public Sector

The adoption of MAS in the public sector, especially in developing countries, faces significant challenges, including resource constraints, infrastructure limitations, expertise gaps, and the need to bridge the research-to-implementation divide. However, despite these obstacles, more than hundreds of MAS-derived varieties of agricultural and horticultural crops were released from public sectors across 22 countries [11].

One of the most notable achievements in applying MAS has been its success in introgressing and/or combining major-effect genes [12]. This approach has resulted in the commercial release of numerous crop varieties with enhanced resistance to biotic stresses in crops such as rice, wheat, barley, sorghum, pearl millet, beans, groundnut, soyabeans, tomato, chillies and cassava. Approximately two-thirds of the identified varieties have been developed using MAS to enhance biotic stress tolerance. Furthermore, effective applications of MAS have yielded crop variants featuring enhanced quality characteristics. Examples include high-quality protein maize, low-amylose rice, and wheat varieties with reduced cadmium content and higher protein content. Public MAS-breeding efforts have also contributed to enhancing complex traits, such as drought and salt tolerance in rice, as well as improving yields in rice and tomato. This section provides a brief overview of crop varieties developed through MAS in India as of 2022.

2.4.2.1. Rice:

A total of 43 rice varieties, developed using Marker-Assisted Selection (MAS), have been released. These varieties are notable for their resistance to various biotic and abiotic stresses [13].

Trait Improved	Genes introgressed	Varieties
Bacterial blight resistance	*xa4, xa13*	PR 124
	xa13, xa21	Improved Pusa Basmati 1, Pusa 1592, Pusa Basmati 1728, Punjab Basmati 3, Pusa Basmati 1718, Punjab Basmati 4, Punjab Basmati 5
	xa4, xa5, xa13, xa21	Improved Lalat, Improved Tapaswini
	xa5, xa13, xa21	Improved Samba Mahsuri, , CR Dhan 800
	xa4, xa13, xa21	PR 123, PR 122, PR 121
	Xa45(t)	PR 127
	Xa33	DRR Dhan 59
Blast resistance	*Pi2* and *Pi54*	Pusa 6 (Pusa 1612), Pusa Basmati 1609
	Pi9	Pusa Basmati 1637
	Pi2	DRR Dhan 51
	Pi1, Pi54 and *Pita*	Pusa Samba 1850
	Xa21, xa13, xa5, Pi2 and *Pi-54*	DRR Dhan 62
Bacterial blight & blast resistance	*Xa21, xa13, Pi2* and *Pi-54*	Pusa Basmati 1847, Pusa Basmati 1885, Pusa Basmati 1886
Bacterial blight resistance & seedling stage salinity tolerance	*Xa21, xa13, xa5* and *qSaltol*	DRR Dhan 58
Bacterial blight resistance & low soil phosphorous tolerance	*Xa21, xa13, xa5* and *qPup1*	DRR Dhan 60
Submergence tolerance	*qSub1*	Swarna Sub 1, Samba Sub 1, CR 1009 Sub 1, Ranjit Sub 1, Bahadur Sub 1, CO 43 Sub 1, IR 64 Sub1, CR Dhan 803 (Trilochan)
Drought tolerance	*qDTY2.2* and *qDTY4.1*	IR 64 Drt1 (DRR Dhan 42)
Submergence & drought tolerance	*qSub1, qDTY2.1* and *qDTY3.1*	DRR Dhan 50, CR Dhan 801, CR Dhan 802 (Subhash)
Herbicide (Imazethapyr) tolerance	*AHAS*	Pusa Basmati 1979, Pusa Basmati 1985

2.4.2.2. Wheat

A total of five rust-resistant wheat varieties are developed through MAS by Punjab Agricultural University under ICAR-AICRP on wheat and Barley [13].

Trait Improved	Genes Introgressed	Varieties
Stripe rust resistance	*Yr15*	PBW 761 (*Unnat* PBW 550), PBW 757
	Yr10	PBW 752
Stripe & leaf rust resistance	*Yr17, Yr40, Lr37* and *Lr57*	PBW 723 (*Unnat* PBW 343)
	Yr40 and *Lr57*	PBW 771

2.4.2.3. Maize

Quality Protein Maize (QPM) is a fortified maize hybrids with twice the amino acids lysine and tryptophan compared to regular maize varieties. A total of ten QPM and provitamin A biofortified maize varieties has been developed from Vivekananda Parvatiya Krishi Anusandhan Sansthan and ICAR-Indian Agricultural Research Institute to alleviate malnutrition [13].

Trait Improved	Genes introgressed	Varieties
Lysine and Tryptophan	*opaque2*	**Vivek QPM9:** lysine (4.19%) tryptophan (0.83%) **Pusa HM4 Improved:** lysine (3.62%) and tryptophan (0.91%) **Pusa HM8 Improved:** lysine (4.18%) and tryptophan (1.06%) **Pusa HM9 Improved:** lysine (2.97%) and tryptophan (0.68%)
Provitamin-A	*crtRB1*	Pusa Vivek QPM9 Improved, Pusa Vivek Hybrid-27 Improved,
	crtRB1 and *lcyE*	Pusa HQPM-5 Improved, Pusa HQPM-7 Improved, Pusa HQPM-1 Improved, Pusa Biofortified Maize Hybrid-1

2.4.2.4. Pearl Millet

A total of two downy mildew resistant hybrids viz., HHB 67 Improved (*qRSg1* and *qRSg4)* and HHB 67 Improved 2 (*qRSg3.1, qRSg4.2* and *qRSg6.1*) with have been released from Haryana Agriculture University under ICAR-AICRP on pearl millet in collaboration with ICRISAT [2,13].

2.4.2.5. Chickpea

A total of six varieties with fusarium wilt resistance and drought tolerance has been released from Indian Institute of Pulse Research, IARI and ICRISAT [13].

Traits Improved	Genes Introgressed	Varieties
Fusarium wilt resistance	*foc2*	IPCMB 19-3 (Samriddhi
Drought tolerance	*QTL* hotspot on LG4	Pusa Chickpea 10216, Pusa Chickpea 4005, IPCL4-14

2.4.2.6. Groundnut

Two varieties *viz.,* Girnar4 and Girnar 5 have been developed by were introgression of *ahFAD2a* and *ahFAD2b* genes to improve the oleic acid content in ICAR-Directorate of Groundnut Research through molecular breeding approaches.

2.4.2.7. Soyabean

A total of six varieties have been released from ICAR-Indian Institute of Soybean Research through molecular breeding approaches[13].

Traits Improved	Genes introgressed	Varieties
Kunitz trypsin inhibitor free	Null allele of *KTi3*	NRC 127, MACSNRC 1667
Less beany flavour	Null allele of *lox2*	NRC 132
Kunitz trypsin inhibitor free & less beany flavour	Null allele of *KTi3* and *lox2*	NRC 142
YMV resistance	*Rymv*	NRCSL 1
Early maturity	Null allele of *E1*	NRC 138

Conclusion

Molecular breeding is not the substitute for conventional breeding; instead, it is a powerful complement that enhances and expedites the breeding process while reducing breeding cycles. Molecular markers play a crucial role by assisting in the selection process. The evolution of molecular marker technology spans over three decades, starting with the identification of the first marker, RFLP (Restriction Fragment Length Polymorphism), and it has reached its zenith with the advent of advanced sequencing technologies.

The recent integration of high-throughput genotyping and high-throughput phenotyping facilities has further elevated the significance of Marker-Assisted Selection (MAS) in plant breeding. This convergence of advanced genetic marker analysis and precise phenotypic evaluation enriches the scientific approach to breeding. It empowers breeders with a more comprehensive and efficient toolkit to develop improved crop varieties, ultimately contributing to global food security and agricultural sustainability.

References

Nadeem MA, Nawaz MA, Shahid MQ, Doğan Y, Comertpay G, Yıldız M *et al*. DNA molecular markers in plant breeding: current status and recent advancements in genomic selection and genome editing. Biotechnology & Biotechnological Equipment. 2018;32(2):261-85.

Singh B D and Singh A K. Marker Assisted Plant Breeding: Principles and Practices. Springer, New Delhi, 2015, 431-450.

Mondini L, Noorani A, Pagnotta MA. Assessing plant genetic diversity by molecular tools. Diversity. 2009;1(1):19-35.

Kumawat G, Kumawat CK, Chandra K, Pandey S, Chand S, Mishra UN *et al*. Insights into marker assisted selection and its applications in plant breeding. In Plant Breeding-Current and Future Views. 2020.

Collard BC, Mackill DJ. Marker-assisted selection: an approach for precision plant breeding in the twenty-first century. Philosophical Transactions of the Royal Society B: Biological Sciences. 2008 ;363(1491):557-72.

Holland JB. Implementation of molecular markers for quantitative traits in breeding programs—challenges and opportunities. In Proceedings of the 4th international crop science congress 2004. Vol. 26:1-13.

Hospital F. Selection in backcross programmes. Philosophical Transactions of the Royal Society B: Biological Sciences. 2005 ;360(1459):1503-11.

Parmar DL. Foreground Selection in Marker Assisted Backcross. Vigyan Varta.2020; 1(8): 8-10.

Eathington SR, Crosbie TM, Edwards MD, Reiter RS, Bull JK. Molecular markers in a commercial breeding program. Crop Science 2007; 154-163.

Meyer R, Ratinger T, Voss-Fels KP. Options for feeding 10 billion people–Plant breeding and innovative agriculture. Report prepared for STOA, the European Parliament Science and Technology Options Assessment Panel, under contract IP. A/STOA/FWC/2008-096/LOT3; 2013.

Vogel B. Marker assisted selection: A biotechnology for plant breeding without genetic engineering. Smart breeding: The next generation. 2014;8.

Varshney RK, Bohra A, Yu J, Graner A, Zhang Q, Sorrells ME. Designing future crops: genomics-assisted breeding comes of age. Trends in Plant Science. 2021;26(6):631-49.

Yadava DK, Hossain F, Choudhury PR, Kumar D, Singh AK, Sharma TR, Mohapatra T. Crop cultivars developed through molecular breeding.2022.1-80.

Chapter-3

Concepts and Applications of Multi-Trait Selection in Plant Breeding

Dhairya V. Makwana[1], Divya S. Patel[2], Mukesh R. Parmar[3] and *Purnima Ray[4]

1,2,3 Ph.D. Student, Dept. of Plant Breeding and Genetics, Navsari Agricultural University, Navsari, Gujarat, India.

4. Ph.D Student, Dept. of Plant Breeding and Genetics, Navsari Agricultural University, Navsari, Gujarat, India.

**Corresponding author: purnimaroy36@gmail.com*

3.1. Introduction

For millennia, plant breeding has been a basic practise in agriculture, with the goal of developing new superior crop varieties to suit the requirements of a growing global population. Breeders have traditionally used single trait selection approaches, in which a particular desirable feature, such as yield, disease resistance, or quality, is emphasised for improvement. This strategy, however, may result in a trade-off between characteristics limiting the overall performance and adaptability of the released cultivars.

Multi trait selection, on the other hand, entails improving numerous qualities at the same time in a single breeding programme. Breeders can produce more balanced crop enhancements that match with the numerous and dynamic difficulties faced in modern agriculture by examining several attributes at once.

Climate change, limited resources, and shifting customer tastes make multi trait selection especially important.

3.1.1. Need for Multi Trait Selection

Because crops must meet many needs, such as yield stability, stress tolerance, and nutritional quality, single trait selection has constraints. Multi trait selection addresses these issues holistically by simultaneously enhancing numerous desired features in agricultural plants.

3.1.2. Objectives of the Chapter

This chapter will look at the concepts and applications of multi-trait selection in plant breeding, with an emphasis on the benefits, problems, and recent breakthroughs that have revolutionised the field. It includes up to date research papers and case examples to demonstrate successful applications.

3.2. Fundamental Concepts of Multi-Trait Selection

3.2.1 Single Trait vs. Multi Trait Selection

Single-trait selection entails focusing on one trait at a time, which frequently results in unwanted side effects on other qualities. Multi trait selection, on the other hand, analyses numerous traits at the same time, taking into account genetic relationships and trade-offs.

3.2.2 Genetic Correlation and Its Importance

There are genetic correlations between distinct qualities, which means that their genes are frequently connected or shared. Understanding genetic connections is critical for multi-trait selection because it allows you to predict how changes in one characteristic will impact others.

3.2.3 Trade-offs in Multi-Trait Selection

Multi trait selection frequently entails trade-offs, in which one feature may be improved at the cost of another. Understanding and managing these trade-offs is critical for breeding programmes to be successful.

3.2.4 Breeding Indices and Selection Strategies

Breeding indices aggregate many features into a single value to help breeders rank and pick superior individuals. Depending on the breeding goals, different selection procedures, such as tandem selection or selection indices, might be used.[2]

3.3. Breeding Schemes for Multiple Traits

Several strategies have been developed to cope with the task of multiple trait selection and integration. To breed for different qualities in crop, breeders must first evaluate which features are most significant, taking into account breeding populations and resources. Tandem selection, independent culling and selection index are three basic ways for breeding or selecting for many features at the same time.[3]

3.3.1. Tandem Selection

It aims to develop a breeding population for many qualities by focusing on one trait at a time for several generations, then switching to another trait for the following breeding cycle or period (serial improvement). A breeder's main concern is knowing how long each trait is selected and at what intensity the trait is picked before switching to another trait. This way of selection is effective when there is no link between the traits or when the relative value of each feature fluctuates over time. For example, if genetic associations between yield and disease resistance do not exist, tandem selection can be employed efficiently to raise disease resistance before yield selection begins.

3.3.2. Independent Culling

It is also known as truncation selection, involves selecting for numerous features from a single population in a certain order in one generation. Breeders must keep in mind the following considerations before using independent culling:

- ❐ Breeders must maintain a large enough population after each level of culling to ensure that there is enough variation left for the traits in subsequent culling;

- ❐ Breeders must use less rigorous culling for the first trait to ensure there is enough variation for an unfavorably associated trait;

- ❐ When genetic correlations are insignificant breeders must keep in mind the order of culling, depending upon the economics of the breeding program and importance of the trait.

As an example, in a greenhouse, culling for rust disease resistance may be done at the seedling stage, and those genotypes that weren't culled at that stage might later be further tested for yield and quality characteristics in the field.

3.3.3. Selection Index

The negative correlation between desirable features is one of the most difficult aspects of selecting for numerous traits. It is critical to overcome the negative correlation between the features in order to produce desired genotypes. Index selection is a strategy for simultaneously increasing qualities when there is a negative association between the traits. The "index," which is a consequence of the different characteristics being selected in index selection, is the single new trait that the breeder creates. Index selection basically involves assessing certain traits according to their economic importance and potential for improvement. Given the idea of aggregate genetic value, index selection is the theoretically most effective strategy for improving crop merit.[4]

The aggregate genetic value of the genotype is its performance based on multiple traits being considered. This aggregate genetic value (H) is given as:

$$H = a_i g_i$$

Where,

ai = economic value or relative value for trait i and

Gi = genetic value for trait i.

3.4. Molecular breeding Schemes for Multi-Trait Selection

In order to develop a specific trait, molecular breeding uses genomics, linkage maps and genetic markers. Marker-assisted selection (MAS), marker assisted backcrossing (MABC), marker-assisted recurrent selection (MARS) and genomic selection or genomic predictions (GS) are a few examples of molecular breeding techniques.[5] Individuals are chosen based on the marker pattern in a process known as "marker assisted selection," which is an indirect phenotypic selection method. [6] [7] [8] [9] [10]

3.4.1. Marker-Assisted Selection (MAS)

Marker assisted selection refers to the use of molecular markers linked to target features to aid in selection. Breeders can indirectly select for numerous traits at the same time by finding marker trait relationships.

3.4.2. Marker-Assisted Backcrossing

The most basic form of MAS is marker-assisted backcrossing. Marker-assisted backcrossing (MABC) refers to a backcross programme helped by molecular markers. It tries to introduce one or a few target genes/QTLs of interest from a donor line (which may be agronomically inferior but includes desired genes for specific traits) into a desired genetic background (which may be agronomically superior). The receiver parent serves as a recurrent parent, with the goal of reducing donor DNA content in succeeding generations through frequent backcrossing to the recurrent parent.

3.4.3. Gene Pyramiding or Gene Stacking

Gene pyramiding is a technique that combines numerous favourable genes from multiple parents into a single genotype. Gene pyramiding is also known as multi trait introgression because genes controlling two or more traits are frequently introgressed into a single recurrent parent. The introgression of multiple QTLs/genes and its effects have been proposed in crop species like wheat, barley, rice, and soybean. [11] [12] [13] [14] [15]

3.4.4. Marker-Assisted Recurrent Selection

In MARS, numerous genomic regions that have complex characteristics are selected and identified using molecular markers to determine the most effective genotype within a single population or among related populations.[5] MARS makes it possible to combine advantageous genes from many genetic origins into a single genetic background. For one cycle of selection, it permits genotypic selection and inter crossing throughout the same crop season .By raising the frequency of advantageous alleles, MARS improved the effectiveness of long term selection in maize.[16] In addition, the genetic gain made possible by MARS in maize was almost twice as great as that made possible by phenotypic selection (PS) in some reference populations.[17] [18]

3.4.5. Genomic Selection to Improve Multiple Traits

Genomic selection also known as genome-wide selection, is a particular kind of marker-assisted selection in which selection is based on data from markers distributed throughout the whole genome. To forecast an individual plant's performance before field testing, statistical modeling and revolutionary bioinformatics methods are used. The potential for genomic selection in line breeding to improve the selection for important agronomic traits like yield has been demonstrated in a number of studies in cereals and legumes. [19] [20] [21] [22]

A general overview of various breeding methodologies, including conventional and molecular, to breed for multiple traits and genes is given in Fig. 1.

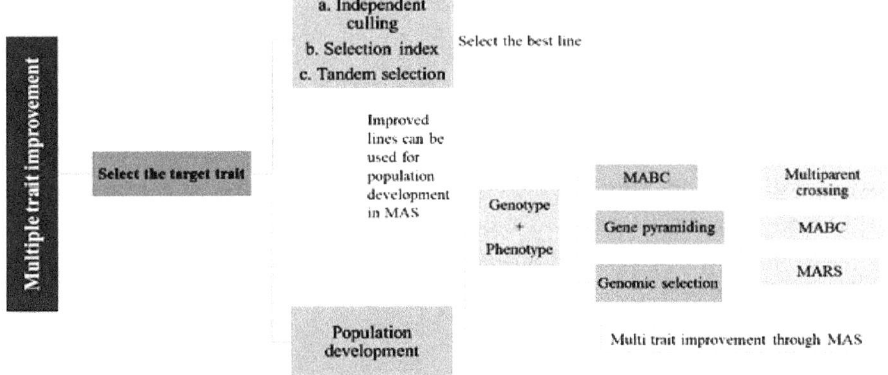

Fig.3.1: Multiple Trait Improvement Through Various Breeding Methodologies

3.5. Statistical Methods for Multi-Trait Selection

Multi-trait statistical approaches are critical in analysing and making educated judgements based on data from various traits. These strategies assist breeders in identifying individuals with favourable trait combinations and accurately estimating breeding values. Below are some common statistical methods used in multi-trait selection, along with appropriate references for each content:

3.5.1 Multivariate Analysis

- ❑ Multivariate analysis involves analyzing multiple traits simultaneously using statistical techniques like multivariate regression, principal component analysis (PCA) and canonical correlation analysis (CCA).

- ❑ Multivariate regression allows breeders to estimate the genetic relationship between traits and identify which traits contribute most to the overall variation.

- ❑ PCA and CCA help in summarizing the information across traits and identify patterns of variation that may not be evident in individual trait analyses[23].

3.5.2. Mixed Model Approaches

- ❑ These are widely used in plant and animal breeding due to their ability to handle complex data structures and account for various sources of variability, such as genetic, environmental and random effects.

- ❑ In multi trait selection, mixed models are extended to include genetic covariance matrices to model the genetic relationships between traits.

- ❑ Best Linear Unbiased Prediction (BLUP) and restricted maximum likelihood (REML) are common methods used in mixed model approaches. [24] [25]

3.5.3. Bayesian Methods

- ❑ Bayesian methods are increasingly used in multi trait selection due to their ability to incorporate prior information, flexibility in modelling complex genetic architectures and providing credible intervals for parameter estimates.

- ❑ Bayesian multiple trait models can estimate the genetic correlations between traits and quantify uncertainties associated with parameter estimates. [26]

3.5.4. Selection Index Theory

- ❑ Selection index theory is a statistical framework that combines information from multiple traits into a single selection index based on economic weights and estimated genetic parameters.

- ❑ The theory allows breeders to optimize selection decisions to achieve the best genetic progress across multiple traits simultaneously. [27]

These statistical methods provide powerful tools for breeders to analyze multi trait data and make effective selection decisions, leading to improved breeding outcomes across multiple important traits. Each method has its advantages and limitations, and their choice depends on the specific characteristics of the breeding program and the available data.

3.6. Advantages of Multi-Trait Selection

Multi trait selection offers several advantages over the traditional single trait approach:

- ❏ **Enhanced Genetic Gain:** Selecting for multiple traits simultaneously allows breeders to make progress in multiple directions, leading to enhanced genetic gain per breeding cycle.

- ❏ **Targeting Complex Traits:** Multiple genes and environmental variables control many desirable characteristics, including the ability to withstand drought, disease resistance and nutritional content. The use of multiple traits in selection helps to improve and target these complex traits.

- ❏ **Reduced Time and Resources:** By assessing multiple traits at once, breeders can reduce the time and resources required for breeding programs, accelerating the development of improved varieties.

- ❏ **Addressing Trade-Offs:** Some traits may have trade-offs, where improvement in one trait could negatively impact another. Multi trait selection helps strike a balance between conflicting objectives.

3.7. Applications of Multi Trait Selection

- ❏ **Yield and Quality Enhancement:** In many cases, the pursuit of high crop yield must be balanced with maintaining product quality. Multi trait selection can target both yield related traits (such as grain weight, ear length, and number of seeds per pod) and quality attributes (like protein content, oil composition, and mineral content). For example, in maize breeding, a combination of yield components and nutritional traits has been targeted using multi trait selection to develop hybrids that offer improved nutritional value without compromising yield potential. [28]

- ❏ **Disease Resistance and Stress Tolerance:** Breeding for disease resistance and stress tolerance often involves multiple interacting traits. Multi trait selection enables breeders to enhance a plant's ability to withstand biotic and abiotic stresses simultaneously. For instance, in rice breeding, combining traits related to blast disease resistance and submergence tolerance has led to the development of high yielding varieties that can withstand flood prone areas. [29]

- ❏ **Climate Resilience:** With climate change impacting agricultural production, multi trait selection can aid in developing crops that are resilient to changing environmental conditions. Integrating traits like heat tolerance, drought resistance, and efficient water use can help ensure crop productivity under varying climatic scenarios. Wheat breeding programs have utilized multi trait selection to create varieties with improved heat tolerance and water use efficiency. [30]

❏ **Nutritional Enrichment:** Addressing global malnutrition requires the development of crops with enhanced nutritional profiles. Multi trait selection can target a range of nutritional traits, such as increasing micronutrient content (e.g., iron, zinc) in staple crops. Biofortified species developed through multi trait selection are contributing to improved human health in regions with nutrient deficiencies [31]

3.8. Challenges and Limitations

❏ **Phenotyping Complexities in Multi Trait Breeding:** Accurate and high throughput phenotyping is a challenge in multi trait breeding, as some traits are difficult to measure. [32]

❏ **Breeding for Complex Quantitative Traits:** Complex quantitative traits involve the interaction of numerous genes, making breeding more intricate. [33]

❏ **Socio-Economic Challenges and Ethical Considerations:** Multi trait selection should also consider socio-economic implications and ethical aspects associated with genetic modifications.

3.9. Future Perspectives

3.9.1. Emerging Technologies and Their Potential in Multi-Trait Selection

Advancements in high-throughput phenotyping, genome editing, and gene stacking offer promising avenues for advancing multi-trait selection.

3.9.2. Integration of Multi-Omics Approaches

Combining genomic data with other omics information, such as transcriptomics and metabolomics, can further enhance multi trait selection.

3.9.3. Implications for Sustainable Agriculture

The widespread adoption of multi trait selection can play a critical role in achieving sustainable agricultural practices and ensuring food security for future generations.

3.9.4. Importance of Stakeholder Collaboration and Knowledge Sharing

Collaboration between researchers, breeders, policymakers, and farmers is vital for successful implementation and adoption of multi trait selection strategies.

3.9.5. Prospects for Precision Breeding and AI-Driven Strategies

Integration of artificial intelligence and precision breeding techniques holds great potential for accelerating multi-trait selection.

3.10. Conclusion

Multi-trait selection in plant breeding is an effective method for developing superior crop varieties with improved performance across many attributes. Plant breeders can make more informed decisions and speed the production of resilient, productive, and nutritionally enriched crops by evaluating trait correlations and employing modern statistical approaches, contributing to global food security and sustainable agriculture.

To generate efficient cultivars that match consumer needs in difficult agricultural conditions, it is vital to breed for numerous attributes such as yield and quality as well as tolerance to abiotic and biotic challenges. Though difficult, conventional breeding approaches for integrating numerous qualities into one genetic background have significant potential in breeding. To breed for numerous qualities, breeders must prioritise desired traits as well as the availability of resources, cost, and time. Marker assisted selection, marker assisted backcrossing, gene pyramiding, and genomic selection are examples of molecular breeding technologies that can considerably augment conventional breeding and aid in multiple trait or gene integration. Effective genomic resources, such as high-throughput marker identification and development, high density linkage maps and gene identification approaches are always in demand.

References

1. Crossa, J., Pérez-Rodríguez, P., Cuevas, J., Montesinos-López, O., Jarquín, D., De Los Campos, G. et al. Genomic selection in plant breeding: methods, models, and perspectives. Trends in plant science. 2017; 22(11): 961-975.

2. Smith, A. B., Cullis, B. R. and Thompson, R. The analysis of crop cultivar breeding and evaluation trials: an overview of current mixed model approaches. The Journal of Agricultural Science. 2005; 143(6): 449-462.

3. Hazel, L. N., and Lush, J. L. The efficiency of three methods of selection. Journal of Heredity. 1942; 33(11): 393-399.

4. Bernardo, R., What if we knew all the genes for a quantitative trait in hybrid crops? Crop Sci. 2001; 41: 1-4.

5. Ribaut, J. M., de Vicente, M. C. and Delannay, X. Molecular breeding in developing countries: challenges and perspectives. Curr. Opin. Plant Biol. 2010; 13: 213-218.

6. Collard, B. C. Y., Jahufer, M. Z. Z., Brouwer, J. B. and E. C. K Pang. An introduction to markers, quantitative trait loci (QTL) mapping and marker-assisted selection for crop improvement: the basic concepts. Euphytica. 2005; 142: 169-196.

7. Collard, B. C. Y., Mackill, D. J. Marker-assisted selection: an approach for precision plant breeding in the twenty-first century. Philos. Trans. R. Soc. Lond. Ser. B Biol. Sci. 2008; 363: 557-572.

8. Jena, K. K. and Mackill, D. J. Molecular markers and their use in marker-assisted selection in rice. Crop Sci. 2008; 48: 1266-1276.

9. Xu, Y., Crouch and J. H. Marker-assisted selection in plant breeding: from publications to practice. Crop Sci. 2008; 48: 391-407.

10. Gupta, P. K., Kumar, J. and Mir, R. R. Marker-assisted selection as a component of conventional plant breeding. Plant Breed. Rev. 2010; 33: 145-217.

11. Richardson, K. L., Vales, M. I., Kling, J. G., Mundt, C. C. and Hayes, P. M. Pyramiding and dissecting disease resistance QTL to barley stripe rust. Theor. Appl. Genet. 2006; 113: 485-495.

12. Jiang, G. L., Dong, Y., Shi, J. and Ward, R. W. QTL analysis of resistance to Fusarium head blight in the novel wheat germplasm CJ 9306. II. Resistance to deoxynivalenol accumulation and grain yield loss. Theor. Appl. Genet. 2007b; 115: 1043-1052

13. Jiang, G. L., Shi, J. and Ward, R. W. QTL analysis of resistance to Fusarium head blight in the novel wheat germplasm CJ 9306. I. Resistance to fungal spread. Theor. Appl. Genet. 2007a; 116: 3-13.

14. Wang, X., Jiang, G. L., Green, M., Scott, R. A., Hyten, D. L. and Cregan, P. B. Quantitative trait locus analysis of saturated fatty acids in a population of recombinant inbred lines of soybean. Mol. Breed. 2012. https://doi.org/10.1007/s11032-012-9704-0.

15. Luo, Y., Ma, T., Zhang, A., Ong, K. H., Li, Z., Yang, J. et al. Marker-assisted breeding of the rice restorer line Wanhui 6725 for disease resistance, submergence tolerance and aromatic fragrance. Rice. 2016. https://doi.org/10.1186/s12284-016-0139-9.

16. Johnson, R. Marker assisted selection. In: Jannick, J. (Ed.), Plant Breeding Review. vol. 24(1), 2004. 293-309.

17. Eathington, S. R. In: Practical applications of molecular technology in the development of commercial maize hybrids. Proceedings of the 60th Annual Corn and Sorghum Seed Research Conferences. American Seed Trade Association, Washington, DC, 2005.

18. Crosbie, T. M., Eathington, S. R., Johnson, G. R. and Edwards, M. Plant breeding: past, present and future. In: Lamkey, K.R., Lee, M. (Eds.), Plant Breeding: The Arnel R. Hallauer International Symposium. Blackwell Publishing, Oxford, UK, 2006, pp. 3-50.

19. Jarquín, D., Crossa, J. and Lacaze, X. A reaction norm model for genomic selection using high-dimensional genomic and environmental data. Theor. Appl. Genet. 2014; 127: 595-607.

20. Tayeh, N., Klein, A., Le Paslier and M. C. Genomic prediction in pea: effect of marker density and training population size and composition on prediction accuracy. Front Plant Sci. 2015; 6: 1-11.

21. He, S., Schulthess, A. W. and Mirdita, V. Genomic selection in a commercial winter wheat population. Theor. Appl. Genet. 2016; 129: 641-651.

22. Michel, S., Ametz, C. and Gungor, H. Genomic selection across multiple breeding cycles in applied bread wheat breeding. Theor. Appl. Genet. 2016; 129: 1179-1189.

23. Piepho, H. P. and Möhring, J. Computing heritability and selection response from unbalanced plant breeding trials. Genetics. 2007; 177(4): 1881-1888.

24. Henderson, C. R. Best linear unbiased estimation and prediction under a selection model. Biometrics. 1975; 31(2): 423-447.

25. Piepho, H. P. A mixed model approach to mapping quantitative trait loci in barley on the basis of multiple environment data. Genetics. 2000; 156(4): 2043-2050.

26. Gianola, D., and Fernando, R. L. Bayesian methods in animal breeding theory. Journal of Animal Science. 1986; 63(1): 217-244.

27. Hazel, L. N. The Genetic Basis for Constructing Selection Indices. Genetics. 1943; 28(6): 476-490.

28. Holland, J. B. Genetic architecture of complex traits in plants. Current Opinion in Plant Biology. 2007; 10(2): 156-161.

29. Septiningsih, E. M., Prasetiyono, J., Lubis, E., Tai, T. H., Tjubaryat, T., Moeljopawiro, S. et al. Identification of quantitative trait loci for yield and yield components in an advanced backcross population derived from the *Oryza sativa* variety IR64 and the wild relative *O. rufipogon*. Theoretical and Applied Genetics. 2003; 107(8): 1419-1432.

30. Reynolds, M. P., Pierre, C. S., Saad, A. S., Vargas, M. and Condon, A.G. Evaluating potential genetic gains in wheat associated with stress-adaptive trait expression in elite genetic resources under drought and heat stress. Crop Science. 2007; 47(3): 172-189.

31. Bouis, H. E and Saltzman, A. Improving nutrition through biofortification: A review of evidence from Harvest Plus, 2003 through 2016. Global Food Security. 2017; 12: 49-58.

32. Araus, J. L., and Cairns, J. E. Field high-throughput phenotyping: the new crop breeding frontier. Trends in Plant Science. 2014; 19(1): 52-61.

33. Bandillo, N., Raghavan, C., Muyco, P. A., Sevilla, M. A., Lobina, I. T., Dilla-Ermita, C. J. et al. Multi-parent advanced generation inter-cross (MAGIC) populations in rice: progress and potential for genetics research and breeding. Rice. 2013; 6(1): 11.

Chapter–4

From Theory to Practice: Advancements in Genomic Selection and Predictive Breeding

Isha K. Mendapara[1]*****, **Rukhsar Bamji**[2]

1   Ph.D. (Agri.) GPB, Department of Genetics and Plant Breeding, N. M. College of Agriculture, NAU.

2   Ph.D. (Agri.) GPB, Department of Genetics and Plant Breeding, N. M. College of Agriculture, NAU.

**Corresponding author: ikmendapara@gmail.com*

4.1. Introduction

The world faces an ever-growing challenge in the 21st century: feeding a global population that is projected to reach nearly 10 billion by 2050, while simultaneously addressing the environmental, economic, and societal constraints placed upon modern agriculture. In this complex landscape, the science of plant breeding stands as a linchpin for addressing these challenges, ensuring food security, and fostering sustainable agricultural practices.

Traditionally, plant breeding was a meticulous, time-consuming process that relied heavily on phenotypic observations and the painstaking selection of individual plants with desired traits over multiple generations. However, advancements in marker techniques have played a pivotal role in enhancing our understanding of genetics, enabling more precise breeding programs, and accelerating scientific research through linkage mapping, QTL mapping, Marker Assisted Selection (MAS) and association mapping . These are all valuable tools in the field of genomic selection and plant breeding. However, they also come with certain drawbacks and limitations. MAS relies on a limited number of markers scattered throughout the genome which inadvertently reduce genetic diversity if breeders overly rely on a small set of markers, potentially leading to inbreeding and vulnerability to diseases and environmental stress. Thus, low resolution makes it difficult to precisely pinpoint the genes responsible for complex traits. Moreover MAS along with Marker Assisted Back Crossing (MABC) and Marker assisted Gene Pyramiding (MAGP) focus on improvement of introgressed QTLs [2]. Linkage mapping typically requires controlled crosses and the generation of segregating populations, which can be labor-intensive and time-consuming. This limits the size of populations, reducing statistical power and it is most suitable for the mapping of single gene Mendelian traits. Complex traits controlled by multiple genes are challenging to dissect through linkage mapping alone. Linkage maps provide information about the order and relative distances of markers but do not capture haplotype information. This limits their ability to identify closely linked markers that collectively influence a trait [3]. QTL mapping is most effective for traits controlled by a relatively small number of major genes. For traits influenced by numerous genes with small effects, QTL mapping may have limited resolution and sensitivity [4]. Association mapping is sensitive to population structure, which can lead to false positives or false negatives if not properly accounted for. It requires carefully designed populations and statistical corrections. It can identify marker-trait associations, it does not establish causality and unable to explain all the variance of the traits [5].

In practice, these drawbacks should not discourage the use of MAS, linkage mapping, QTL mapping or association mapping but rather inform breeders and researchers about their limitations. However, a better method would be scoring many markers at low cost which capitalize simultaneous estimation of all marker loci rather than seeking single marker associated with single large effect locus or QTLs. Genomic selection, which incorporates information from all markers across the genome, has emerged as a promising approach that can overcome many of these limitations by integrating data from various sources and using sophisticated prediction models [6]. This approach can hasten the development of superior crop varieties with increased efficiency and precision. These advancements have transformed genetics and genomics, making it possible to address complex biological questions, accelerate breeding programs, and ultimately improve our understanding of life's molecular processes. They continue to evolve, with ongoing research and development aimed at refining existing methods and developing new ones to push the boundaries of genetic research and applications.

4.2. Theoretical Foundations

A new selection approach termed genomic selection (GS) has been proposed that can facilitate selection for targeted traits by calculating an individual's net genetic merit utilizing the impacts of dense markers dispersed across the genome

[7].Genomic selection utilizes two different population to serve its purpose: a training set and validation set or candidate set. Genotyping and phenotyping of training population will help building a statistical model which is validated through genotyping of untested individuals of candidate set to estimate the individual effect of each marker, and the additive sum of all marker effects is utilized to calculate the GEBV of each individual [5]. Breeding value of any individual represent the expected phenotype of its progeny. Conventionally, it was estimated through half-sib progeny by estimating GCA and additive gene actions. Considering decline genotyping cost, stationary or snowballing phenotyping cost alongwith early generation selection of individuals, genomic selection is transmuting both animal and plant breeding strategies [5].

4.3. Phenotypic *vs.* Genomic Selection

Classical breeding has progressed substantially in the last century, making significant contributions to crop improvement by developing high yielding and nutrient responsive varieties. Furthermore, conventional breeding is dependent on phenotypic selection to select better parents for crossing or generation advancement and is less successful for low heritable multigenic quantitative characteristics which are significantly influenced by environment and G × E interaction. Additionally, these methods are hampered by the fact that they are time consuming, laborious, requires huge amount of land, are inefficient in terms of cost, and are less exact and dependable, necessitating the creation of intermediate, fast, and efficient selection procedures for the production of high yielding and climate resilient crop varieties [8].

To overcome these problems, GS a new technique based on reduced phenotyping and selection based on marker/genotypic profile were proposed [7]. GS creates the prediction model by combining genotypic and phenotypic data of training population, which is then utilized to construct GEBV for all individuals of breeding population based on their genotypic data [9]. Selection of new breeding parents are based on the estimated GEBV, which leads to reduced breeding cycle time. GS has a number of advantages over PS in that it saves time, money, and produces more reliable results while also being environmentally insensitive [10,11]. This enables the faster development of improved varieties of crop species compared to conventional or phenotypic selection.

4.4. Marker Development and Marker density

Genomic selection is an improved version of MAS. GS opens up a new study area for molecular breeding because the objectives of selection are no longer restricted to qualities determined by a few major genes [12]. There needs tobe extensive genotyping at many loci in order to undertake crop genomic prediction. The evaluation of QTL where the regions of the genome related with phenotypes was made possible by the introduction of molecular genetic markers *i.e.,* SSR, SNPs, InDels etc. [13]. The effects of each marker on a trait are predicted by genomic selection using a prediction equation based on genome-wide DNA markers. GS is an adaptation of the MAS method that makes use of SNP markers. The aim of genomic selection (GS) is to use linkage disequilibrium between high density markers and quantitative trait loci (QTLs) across the genome to estimate breeding value in crop improvement programs. In order to put GS into practice, it is first necessary to assess the effects of high density single nucleotide polymorphisms (HD-SNPs) using the genotype and phenotype data of individuals for a quantitative character (training) [7]. The estimated marker effects are then applied to the marker genotypes of selection candidates to determine their GEBVs. By allowing calculation of breeding values without the need for phenotypic data, GS has the potential to select candidates for the upcoming breeding cycle based on individual's GEBVs obtained from genome-wide markers [7][14].

Genomic selection is different from existing MAS techniques, because instead of only employing markers that have a predefined significant association with a trait, all markers are used to estimate breeding value of each genotype. Subsequently, to maximize the number of QTLs in LD with at least one marker, dense marker coverage is required. Thus, resulting in increase in the number of QTLs whose effects will be captured by markers [15]. However, upto a limit, marker density enhances GEBV accuracy; thereafter, no gain is observed. Moreover, low density and evenly spaced markers may forecast GEBVs with less precision. Thus, sufficient marker information is required for success of GS strategy.

4.5. GS in Era of NGS

Power of GS vastly hinge on type and number of markers utilized as well as its distribution on the genome, as the focus is not a single trait/QTL. Genome sequencing/resequencing is proven as the definitive tactic to identify polymorphism in any crop [8]. Genome sequencing became more popular after introduction of Next-generation sequencing (NGS) technologies i.e. whole genome sequencing (WGS), whole genome resequencing (WGRS), whole exome sequencing (WES), targeting sequencing, genotyping by sequencing (GBS), transcriptome sequencing etc. due to reduced time and cost with escalated precision of sequencing [16][17]. Initially, Sanger sequencing used for sequencing of whole genome was costly and time consuming with limited application in target gene discovery and was limited to sequencing of model organisms. In contrary, combination of long read third and fourth generation sequencing platforms and short-read second generation sequencing platforms (*i.e.* Illumina) have overcome the limitation of population size and wrong base calling by sequencing many genome at a time with reduced cost and time with increased precision [17].

Availability of WGS have caused paradigm shift from fragment length polymorphism to sequence-based polymorphism (SSR, SNPs etc.) identification to accelerate marker information practice for escalating number of informative markers. NGS became the powerful genomics assisted breeding tool as it avails vast DNA sequence polymorphism within a shorter time frame. However, sequencing of whole genome of large mapping population and various germplasm lines is rather pricey and not practical but the ongoing revolution in NGS technology may reduce this cost in near future which will lead to discovery of large number of markers to meet the need of plant breeding methods including genomic selection. Until then, feasible alternative for large scale marker discovery is targeted sequencing.

Simultaneous SNP discovery, sequencing and genotyping is possible through various NGS methods. Most of which relies on restriction digestion to reduce genome complexities. However, rather than restriction digestion based length polymorphism, developed markers are either SNPs or structural variants. Complexity reduction is required to avoid repetitive region or junk DNA in eukaryotic genome to target lower copy/ unique region by digestion through methylation sensitive restriction enzymes *i.e.* reduced-representation sequencing, including reduced-representation libraries (RRLs) and complexity reduction of polymorphic sequences (CRoPS); restriction site associated DNA sequencing (RAD-seq) and low coverage genotyping, genotyping by sequencing (GBS) [18]. RRLs and CRoPS involve sampling and sequencing of small share of genome-wide region and identify the polymorphism from subset of genome [18]. RRL in its simplest form, only the ends of restriction fragments are sequenced, however it can be modified to sequence entire fragments. When reference genome sequence is available, reads from RRL are aligned to reference for SNP discovery. If reference genome is not available, then SNP calling is difficult as it requires *de novo* assembly of sequence [18]. CRoPS adapts amplification of AFLP fragments to enable sequencing on the Roche Genome Sequencer platform [18]. RAD-seq involve sequencing of region surrounded by essentially all the restriction sites for particular restriction enzyme throughout the genome, irrespective of restriction fragments length. RAD-seq being the effective SNP discovery approach than RRLs and CRoPS for many cases due to higher widespread marker discovery across the genome [18].

Above mentioned genotyping methods involve marker discovery through higher coverage in reduced targeted genomic region to allow genotyping of many individuals. However, alternative to this method is to target many marker regions for sequencing at low coverage. Genotyping by sequencing (GBS) approach is based on this idea which follows digestion of DNA with methylation sensitive frequent cutter and sequencing of end of all the digested fragments without size selection [19]. GBS allows discovery of huge number of SNPs with lower per sample cost in larger number of individuals at a time. At present, GBS is the widely used method due to cost-effectiveness, multiplexed nature for genotyping of large population, RILs, diverse germplasm lines for SNP genotyping, trait mapping and genomic estimation of breeding values. Utilizing genomic selection *via* GBS method represents a significant enhancement to conventional crop improvement techniques, playing a crucial role in advancing genomics-assisted breeding within the realm of commercial crops [20].

4.6. GS and Pangenomics

Inadequacy of single reference genome led to idea of building a complete reference genome describing complete genetic diversity of a species. Thus, branch of pangenomics evolved which combines genomic information of a species [21]. However, in spite of wider diversity represented by pangenome, they have not extensively been used for genomic selection till date. Assembling of pangenome also include wild relatives of concerned species. Thus, it also captures the novel variation which is not present in cultivated varieties. Utilizing only single reference genome, breeders will be unable to identify the phenotype which represent such novel variation as well as markers on large presence/absence variation region. Thus, amalgamation of pangenome in GS could be effectual by identification all possible genome-wide associations.

4.7. Statistical Models used for Genomic Prediction

In breeding programes, statistical approaches for GS are need to estimate multiple marker effects from a small number of phenotypes at the same time. Several statistical methods for genomic prediction and GS implementation have been created. In the evaluation of model performance, GEBV accuracy is defined precisely as the Pearson correlation between the GEBV and the true breeding value (TBV) [15]. When selecting on the GEBV, accuracy is defined as $R = ir_A$, where R is the response, i is the selection intensity, r is the accuracy defined above, and A is the square root of the additive genetic variance of TBV [22]. The models briefly discussed here are;

- ☐ **Bayesian Regression Model:** It uses maximum likelihood approaches to efficiently compute RR-BLUP parameters under the simple assumption of equal and fixed marker effect variances. In this, marker variance is treated more credibly by presumptuous specified prior distribution. Two types of prior distribution for marker variance have been proposed by [7]

  - ☆ **Bayes A:** uses an inverted chi-square to regress the marker variance towards zero and marker effects are > 0.

  - ☆ **Bayes B:** GEBV does not decline with an increase in marker density and some marker effects can be = 0.

- ☐ **Ridge Regression BLUP (RR-BLUP):** All marker effects for GS can be estimated simultaneously using the ridge regression BLUP (RR-BLUP) method [7][23]. Ridge regression reduces all marker effects towards zero rather than classifying markers as significant or having no effect method [23]. It is superior to SR (Stepwise Regression). It has the limitation of equally treating all the effects which is rather unrealistic.

❑ **Stepwise Regression (SR):** Traditional MAS treats marker effects as fixed, necessitating SR techniques that circumvent the lack of degrees of freedom issue by fitting markers separately or in small groups. SR chooses the most significant markers based on arbitrary significant thresholds, and the effect of non-significant markers is set to zero [24]. It estimates the effect of significant markers utilizing multiple regression. It results in low GEBVs accuracy due to limited detection of QTLs.

The major limitation of these classical models is that it does not take into account for non-additive effects such as epistatic and epigenetic effects as well as interaction among genotypes and genotypic and environmental interactions [25][26]. Thus, some novel models such as random forest and machine learning algorithms have been adapted for genomic selection to capture non-linear relationship between marker and traits. Moreover, deep learning models such as deep neural network genomic prediction (DNNGP) have shown promise in genomic prediction by capturing intricate patterns through integration of multi-omics data, but it may require large datasets and substantial computational resources [27].

4.8. GS and Phenomics: Interplay of Genomics and Enviromics

The crucial yet weaken link of genomic selection is phenotyping, as it requires more time and cost to phenotype the large population size accurately. Phenomics is the holistic approach for measurement of various phenotypic traits, or observable characteristics of plants. These traits can include aspects such as growth patterns, disease resistance, yield, morphology, and many others. The goal of phenomics is to quantify and analyse such traits on a large scale, often using advanced technologies like remote sensing, 3D imaging, and high-throughput phenotyping platforms [28]. By collecting extensive data on the phenotypic variations among plants along with genotyping data, genomic selection can enhance accuracy of plant selection for targeted traits. Hence, the primary objective of employing a high-throughput phenotype (HTP) is to lower the per plot data expenditure while enhancing early-season prediction accuracy. This is achieved by leveraging secondary phenotypes that exhibit a high heritability and a close correlation with the targeted selection phenotypes [29].

Since 1960s, G × E interactions[30] are proven fact which suggested use of environmental data along with genomics to predict performance of genotypes across various environments. However, limited environmental factors (temperature, precipitation, radiation) at certain crop growth stage are taken into account by genomic selection models [29]. There is still scope to utilize large scale envirotyping data to train GS models for prediction of various growing conditions by assembling historic enviromic data [29]. However, efforts are going on to link interplay of genomic, phenomic and enviromic data by studying Enviromic + Genomic prediction (E-GP) model for accurate prediction of breeding values [31].

Evidence suggests that prediction accuracy can be enhanced through multi-trait analysis, particularly when genetic and residual correlations are factored into the modelling process. Emerging genomic models, which account for multiple traits and environments, as well as interactions of target trait with environment, genotype as well as G × E, hold substantial promise for harnessing correlations among diverse variables and distinguishing between their effects [29]. Integration of such multi-trait whole genome prediction models of GBLUP (Genomic Best Linear Unbiased Prediction) which is an extension of BLUP that incorporates genomic information along with environmental interactions will improve efficiency of GS [29]. Moreover, Bayesian multi-trait multi-environment (BMTME) models simultaneously consider multiple traits and multiple environments when making predictions based on genomic data. These models leverage Bayesian statistical principles to integrate genetic information, trait data, and environmental factors for more accurate and robust predictions of an individual's or a genotype's performance useful in complex breeding scenarios where multiple traits need improvement and performance of genotypes needs to be evaluated across diverse environmental conditions. in different environments [32].

4.9. Implications of GS in Plant Breeding

In order to tackle the well documented difficulties of food and nutrition security, there is an urgent need to deploy innovative technology to speed up plant breeding advances. These methods can be integrated with traditional phenotypic breeding programs or used to help redesign existing phenotypic breeding pipelines with the view to select better-adapted genotypes in a time and cost efficient manner in order to offer significant advances in breeding pipeline efficiently [29]. Genomic selection is one of the adaptable crop breeding selection techniques based on gene-sequencing technologies. GS being an advanced form of marker-assisted breeding (MAB) ensures that each quantitative trait locus (QTL) is in linkage disequilibrium (LD) with at least one genetic marker which selects for desired trait. GS has been discovered as a possible strategy for improving complicated trait genetics while significantly reducing breeding cycles. Breeding initiatives must begin with early fundamental core parents (training populations) that accurately reflect the genetic diversity of the current offspring and conform as closely as feasible to the testing population(s). These foundation parents should be broadly phenotyped in various target groups and genotyped using high-density marker systems. GS allows for rapid crop improvement without the requirement for detailed investigation of individual loci. Many studies have revealed enormous prospects for GS to boost genetic gain in various crops.

The improvement of traits which are governed by minor genes through conventional breeding and MAS has not produced the desired outcomes in a world with a growing human population [8]. In this regard, GS offers new options for boosting the effectiveness of plant breeding programs by potentially fixing all the genetic variation and efficiently selecting the individuals with high breeding values. Efforts have been made for trait specific improvement in many crops through genomic selection (Table 1).

Table 4.1: Genomic Selection (GS) Efforts Achieved for Various Traits in Different Crops

S. No.	Crop	Genotyping method	Traits	Model	Reference
1	Rice	DAr Tseq	Grain yield, plant height	G-BLUP, RR-BLUP	[33]
		GBS	Grain yield, flowering time	RR-BLUP	[34]
		GBS	Nitrogen balance index	G-BLUP, Bayes B	[35]
		GBS	Blast resistance	RR-BLUP, G-BLUP	[36]
2	Wheat	GBS	Stem rust resistance	G-BLUP B	[37]
		GBS	Grain yield, protein content and protein yield	RR-BLUP	[38]
		GBS	Grain yield, *Fusarium* head blight resistance, softness equivalence and flour yield	BLUP	[39]
		GBS	Heat and drought stress	G-BLUP	[40]
		Infinium iSelect 9 K	Yellow rust resistance	G-BLUP, BRR	[41]
3	Maize	GBS	Grain yield, anthesis date, anthesis-silking interval	RR-BLUP	[42]
		GBS	Drought stress	G-BLUP	[43]
		KASP PCR	Striga resistance	BLUP	[44]
		DArTseq	Ear rot disease resistance	RR-BLUP	[45]
4	Soybean	GBS	Yield and other agronomic traits	G-BLUP	[46]
5	Alfalfa	GBS	Biomass yield	BLUP	[47]
6	Chickpea	Whole-genome re-sequencing	Drought tolerance	RRBLUP, Bayesian	[48]
7	Common bean	GBS	Cooking time	Exome capture	[49]
8	Canola	DArTseq	Flowering time	RR-BLUP	[50]
9	Groundnut	Affymetrix GeneTitan®	Yield, protein, rust resistance	Bayesian	[51]
10	Sunflower	GBS	Oil content	BLUP	[52]

4.10. Genomic Selection in Breeding: Challenges Ahead

Despite its considerable potential, implementation of genomic selection in plant breeding faces several challenges. Genomic selection relies on high quality genotypic and phenotypic data. Obtaining large and accurate datasets for both genotyping and phenotyping can be costly and time consuming, especially for certain crops or traits. Moreover, the cost of genotyping can be a significant barrier, particularly for crops with large and complex genomes. Reducing genotyping costs while maintaining data quality is an ongoing challenge. Obtaining precise and consistent phenotypic data across diverse environments can be challenging. High -throughput phenotyping technologies are continually evolving but may not be readily available or affordable for all crops and traits. Some crops have limited genomic diversity, which can reduce the effectiveness of genomic selection. Strategies for introducing and capturing new genetic diversity through pangenome are needed to be applied for crop improvement.

Integrating data from different sources, such as genomics, phenomics, and environmental data, can be complex. Developing effective data management and analysis pipelines is essential. Based on the type of data, genetic architecture of traits and computational resources, selection of the most appropriate genomic selection model should be done. Such models need to account for genotype-environment interactions, as the performance of genotypes can vary across different growing conditions. After estimation of GEBVs, validating the accuracy of genomic predictions is crucial. Accurate validation may require extensive testing in real breeding programs, which can be resource intensive. For traits influenced by numerous genes or with complex genetic architectures, such as disease resistance or abiotic stress tolerance, predicting performance accurately remains a challenge which can be overcome by using multi-trait multi-environment genomic prediction models.

Implementing genomic selection may raise ethical concerns related to genetic modification or intellectual property rights, which can vary by region and crop. Moreover, many breeding programs may lack the expertise and infrastructure needed to adopt genomic selection. Thus, capacity building and training are essential to ensure successful implementation. Genomic selection is a long term strategy that requires ongoing investment in both technology and expertise, some breeding programs may struggle to maintain this commitment.

Despite these challenges, genomic selection holds great promise for accelerating plant breeding by enabling more efficient and precise selection of superior genotypes. Ongoing research and collaboration between scientists, breeders, and policymakers are essential to address these challenges and fully realize the potential of genomic selection in plant breeding.

4.11 Conclusion

In conclusion, genomic selection has emerged as a transformative force in the field of plant breeding. As we navigate the era of NGS and embrace the power of pangenomics, GS continues to evolve, offering unprecedented opportunities and addressing age old challenges in breeding. The application of GS in plant breeding represents a paradigm shift, allowing breeders to harness the full potential of genetic information. Through the integration of genomics, phenomics, and environmental data, GS empowers breeders to make more informed and precise breeding decisions, accelerating the development of improved crop varieties. As we venture into the future, it is clear that GS is not just a tool but a dynamic and essential component of modern plant breeding. The challenges faced, from data quality and quantity to model refinement and genotype-environment interactions, are substantial but surmountable with continued dedication, collaboration, and innovation. The road ahead may be demanding, but the promise of genomic selection is immense, and it is a journey well worth undertaking to shape the future of plant breeding and ensure food security for generations to come.

References

[1]Hasan N, Choudhary S, Naaz N, Sharma N and Laskar, RA. Recent advancements in molecular marker-assisted selection and applications in plant breeding programmes. Journal of Genetic Engineering and Biotechnology. 2021; 19(1):128.

[2]Ahmar S, Gill RA, Jung, KH, Faheem A, Qasim MU, Mubeen, M et al. Conventional and Molecular Techniques from Simple Breeding to Speed Breeding in Crop Plants: Recent Advances and Future Outlook. International Journal of Molecular Science. 2020; 21(7): 2590

[3]Chevalier, FD, Valentim, CL, LoVerde, PT and Anderson, TJ. Efficient linkage mapping using exome capture and extreme QTL in schistosome parasites. BMC Genomics, 2014; 15(1): 617.

[4]Raj, SRG and Nadarajah K. QTL and Candidate Genes: Techniques and Advancement in Abiotic Stress Resistance Breeding of Major Cereals. International Journal of Molecular Sciences. 2022; 24(1): 6.

[5]Jannink, JL, Lorenz AJ and Iwata H. Genomic selection in plant breeding: from theory to practice. Briefings in Functional Genomics. 2010; 9(2): 166-177.

[6]Budhlakoti N, Kushwaha AK, Rai A, Chaturvedi KK, Kumar A, Pradhan AK et al. Genomic Selection: A tool for accelerating the efficiency of molecular breeding for development of climate-resilient crops. Frontiers in Genetics. 2022; 13: 832153.

[7]Meuwissen TH, Hayes BJ and Goddard ME. Prediction of total genetic value using genome-wide dense marker maps. Genetics. 2001; 157:1819-1829.

[8]Bhat, JA, Ali S, Salgotra RK, Mir ZA, Dutta S, Jadon V et al. Genomic selection in the era of next generation sequencing for complex traits in plant breeding. Frontiers in Genetics. 2016; 7: 221.

[9]Poland J, Endelman J, Dawson J, Rutkoski J, Wu S and Manes Y. Genomic selection in wheat breeding using genotyping-by-sequencing. Plant Genome. 2012; 5: 1-11.

[10]Rutkoski JE, Heffner EL and Sorrells ME. Genomic selection for durable stem rust resistance in wheat. Euphytica. 2011; 179(1): 161-173.

[11] Desta ZA and Ortiz R. Genomic selection: Genome-wide prediction in plant improvement. Trends in Plant Science. 2014; 19(9): 592-601.

[12]Wang X, Xu Y, Hu Z and Xu C. Genomic selection methods for crop improvement: Current status and prospects. The Crop Journal. 2018; 6(4): 330-340.

[13] Geldermann, H. Investigations on inheritance of quantitative characters in animals by gene markers. Theoretical Applied Genetics. 1975; 46(7): 319-330.

[14]Habier D, Fernando RL and Dekker JCM. The impact of genetic relationship information on genome-assisted breeding values. Genetics. 2007; 177(4): 2389-2397.

[15]Heffner EL, Sorrells ME and Jannink J. Genomic selection for crop improvement. Crop Science. 2009; 49(1): 1-12.

[16]Varshney RK, Hoisington DA, Nayak SN and Graner A. Molecular Plant Breeding: Methodology and Achievements. 2009; 283–304.

[17]Pervez MT, Hasnain MJ, Abbas SH, Moustafa MF, Aslam N and Shah SSM. A Comprehensive Review of Performance of Next-Generation Sequencing Platforms. BioMed Research International. 2022; 2022:1–12.

[18]Davey JW, Hohenlohe PA, Etter PD, Boone JQ, Catchen JM and Blaxter ML. Genome-wide genetic marker discovery and genotyping using next-generation sequencing. Nature Reviews Genetics. 2011; 12(7): 499–510.

[19] Elshire RJ, Glaubitz JC, Sun Q, Poland JA, Kawamoto K, Buckler ES et al. A Robust, Simple Genotyping-by-Sequencing (GBS) Approach for High Diversity Species. PLoS ONE. 2011; 6(5): e19379.

[20] Poland J, Endelman J, Dawson J, Rutkoski J, Wu S and Manes Y. Genomic selection in wheat breeding using genotyping-by-sequencing. Plant Genome. 2012; 5(2): 1-11.

[21]Zhao J, Bayer PE, Ruperao P, Saxena RK, Khan AW, Golicz AA, et al. Trait associations in the pangenome of pigeon pea (Cajanus cajan). Plant Biotechnology Journal. 2020; 18(9): 1946–1954.

[22]D. Falconer and T. Mackay, Quantitative genetics. Longman, Harrow, UK. 1996.

[23]Whittaker JC, Thompson R and Denham MC. Marker-assisted selection using ridge regression. Genetics Research. 2000; 75(2): 249-252.

[24]Lande R and Thompson R. Efficiency of marker-assisted selection in the improvement of quantitative traits. Genetics. 1990; 124(3): 743-756.

[25]Varona L, Legarra A, Toro MA and Vitezica ZG. Non-additive Effects in Genomic Selection. Frontiers in Genetics. 2018; 9.

[26]Jiang Y and Reif JC. Modeling Epistasis in Genomic Selection. Genetics. 2015; 201(2): 759–768.

[27]Wang K, Abid MA, Rasheed A, Crossa J, Hearne S and Li H. DNNGP, a deep neural network-based method for genomic prediction using multi-omics data in plants. Molecular Plant. 2023; 16(1): 279–293.

[28] Cobb JN, DeClerck G, Greenberg A, Clark R and McCouch S. Next-generation phenotyping: requirements and strategies for enhancing our understanding of genotype–phenotype relationships and its relevance to crop improvement. Theoretical and Applied Genetics. 2013; 126(4): 867–887.

[29]Crossa J, Fritsche-Neto R, Montesinos-Lopez OA, Costa-Neto G, Dreisigacker S, Montesinos-Lopez A et al. The modern plant breeding triangle: Optimizing the use of genomics, phenomics, and enviromics data. Frontiers in Plant Science. 2021; 12: 651480.

[30]G. H. Freeman and J. M. Perkins. Environmental and genotype-environmental components of variability VIII. Relations between genotypes grown in different environments and measures of these environments. Heredity (Edinb). 1971; 27(1): 15–23.

[31] Rincent R, Kuhn E, Monod H, Oury FX, Rousset M, Allard V et al. Optimization of multi-environment trials for genomic selection based on crop models. Theoretical and Applied Genetics. 2017; 130(8): 1735–1752.

[32]Montesinos-López OA, Montesinos-López A, Luna-Vázquez FJ, Toledo FH, Pérez-Rodríguez P, Lillemo M et al. An R Package for Bayesian Analysis of Multi-environment and Multi-trait Multi-environment Data for Genome-Based Prediction. G3 Genes Genomes Genetics. 2019; 9(5): 1355–1369.

[33]Grenier C, Cao TV, Ospina Y, Quintero C, Châtel MH, Tohme J et al. Accuracy of genomic selection in a rice synthetic population developed for recurrent selection breeding. PLoS ONE. 2015; 10(8): e0136594.

[34]Spindel J, Begum H, Akdemir D, Virk P, Collard B, Redoña E et al. Correction: genomic selection and association mapping in rice (Oryza sativa): Effect of trait genetic architecture, training population composition, marker number and statistical model on accuracy of rice genomic selection in elite, tropical rice breeding lines. Plos Genetics. 2015; 11: e1005350.

[35]Hassen BM, Bartholomé J, Valè G, Cao TV and Ahmadi N. Genomic prediction accounting for genotype by environment interaction offers an effective framework for breeding simultaneously for adaptation to an abiotic stress and performance under normal cropping conditions in rice. Genes. 2018; 8(7): 2319-2332.

[36]Huang M, Balimponya EG, Mgonja EM, McHale LK, Luzi-Kihupi A, Wang GL et al. Use of genomic selection in breeding rice (Oryza sativa L.) for resistance to rice blast (Magnaporthe oryzae). Molecular Breeding. 2019; 39(8):1-16.

[37] Rutkoski JE, Poland JA, Singh RP, Huerta-Espino J, Bhavani S, Barbier H et al. Genomic selection for quantitative adult plant stem rust resistance in wheat. Plant Genome. 2014; 7(3):1–10.

[38]Michel S, Ametz C, Gungor H, Epure D, Grausgruber H, Löschenberger F et al. Genomic selection across multiple breeding cycles in applied bread wheat breeding. Theoretical and Applied Genetics. 2016; 129(6): 1179-1189.

[39]Hoffstetter A, Cabrera A, Huang M and Sneller C. Optimizing training population data and validation of genomic selection for economic traits in soft winter wheat. G3 (Bethesda). 2016; 6(9): 2919-2928.

[40]Crossa J, Jarquin D, Franco J, Pérez-Rodríguez P, Burgueño J, Saint Pierre C et al. Genomic prediction of gene bank wheat landraces. G3 (Bethesda) 2016; 6(7): 1819-1834.

[41]Daetwyler HD, Bansal UK, Bariana HS, Hayden MJ and Hayes BJ. Genomic prediction for rust resistance in diverse wheat landraces. Theoretical and Applied Genetics. 2014; 127(8): 1795-1803.

[42]Crossa J, Beyene Y, Kassa S, Pérez P, Hickey JM, Chen C et al. Genomic prediction in maize breeding populations with genotyping-by-sequencing. G3 (Bethesda). 2013; 3(11): 1903-1926.

[43]Zhang X, Pérez-Rodríguez P, Semagn K, Beyene Y, Babu R, López- Cruz MA et al. Genomic prediction in biparental tropical maize populations in water-stressed and well-water environments using low-density and GBS SNPs. Heredity 2015; 114(3): 291-299.

[44]Badu-Apraku B, Talabi AO, Fakorede MAB, Fasanmade Y, Gedil M, Magorokosho C et al. Yield gains and associated changes in an early yellow bi-parental maize population following genomic selection for striga resistance and drought tolerance. BMC Plant Biology. 2019; 19(1): 129.

[45]Dos Santos JPR, Pires LPM, de Castro Vasconcellos RC, Pereira GS, Von Pinho RG and Balestre M. Genomic selection to resistance to Stenocarpella maydis in maize lines using DArTseq markers. BMC Genetics. 2016; 17(1): 86.

[46] Jarquín D, Kocak K, Posadas L, Hyma K, Jedlicka J. and Graef G. Genotyping by sequencing for genomic prediction in a soybean breeding population. BMC Genomics. 2014; 15(1):740.

[47] Li X, Wei Y, Acharya A, Hansen JL, Crawford JL, Viands DR et al. (2015). Genomic prediction of biomass yield in two selection cycles of a tetraploid alfalfa breeding population. Plant Genome. 2015; 8(2): 1-10.

[48] Li Y, Ruperao P, Batley J, Edwards D, Khan T, Colmer TD et al. Investigating drought tolerance in chickpea using genome-wide association mapping and genomic selection based on whole-genome resequencing data. Frontiers in Plant Science. 2018; 9: 190.

[49] Diaz S, Ariza-Suarez D, Ramdeen R, Aparicio J, Arunachalam N, Hernandez C et al. Genetic architecture and genomic prediction of cooking time in common bean (Phaseolus vulgaris L.). Frontiers in Plant Science. 2021; 11: 2257.

[50] Raman H, Raman R, Coombes N, Song J, Prangnell R and Bandaranayake C. Genome-wide association analyses reveal complex genetic architecture underlying natural variation for flowering time in canola. Plant Cell and Environment. 2015; 39(6): 1228-1239.

[51] Pandey MK, Chaudhari S, Jarquin D, Janila P, Crossa J, Patil SC et al. Genome-based trait prediction in multi-environment breeding trials in groundnut. Theoretical and Applied Genetics. 2020; 133(11): 3101-3117.

[52] Mangin B, Bonnafous F, Blanchet N, Boniface MC, Bret-Mestries E, Carrère S et al. Genomic prediction of sunflower hybrids oil content. Frontiers in Plant Science. 2017; 8: 1633.

Chapter–5

Opportunities and Challenges to Implementing Genomic Selection in Clonally Propagated Crops

Sahana Police Patil[1], Beera Bhavya[1], Yashaswini R.[2] and Adithya P Balakrishnan[1]

1. *Indian Agricultural Research Institute, New Delhi*
2. *University of Agricultural Sciences, Raichur*

5.1. Introduction

Genomic selection represents a revolutionary advancement in the field of plant breeding, offering the potential to revolutionize the development of clonal crops. These crops, which propagate through vegetative parts rather than seeds, are integral to agriculture and encompass economically vital species like grapes, bananas, and citrus fruits. Their improvement has been hindered by their unique reproductive biology. While not an entirely new concept, applying genomic selection to clonal crops presents distinct opportunities and obstacles. The genetic uniformity inherent in clonal crops simplifies certain breeding aspects, yet also makes them susceptible to environmental stresses. Genomic selection, with its advanced genetic analysis tools and statistical models, offers a more efficient means to uncover the genetic basis of complex traits compared to traditional methods. However, integrating genomic selection into clonal crop breeding programs brings forth specific challenges. Tailoring genomic prediction models to account for clonal crop genetic characteristics, including somaclonal variation and ploidy levels, is vital. Reliable phenotypic data is essential, and ethical concerns and intellectual property rights also shape genomic selection in clonal crops. Hence, a thorough exploration of the opportunities and challenges of genomic selection in clonal crops becomes crucial.

5.2. Clonally Propagated Crops

Clonal crops encompass vital root and tuber varieties, diverse forage plants, most fruits, ornamental trees, numerous cut flowers, potted plants, and even forest trees. A crop qualifies as clonally propagated if its cultivation relies on asexually reproduced plant material, irrespective of intraspecies diversity. This includes roots, tubers, stems, corms, and non-meiosis derived seeds. For instance, maize and beans, originally open-pollinated and hybrid crops, would be clonally propagated if propagated via stem cuttings or asexual seeds. Importantly, cross-breeding methods useful for hybrid crops are valuable in developing clonally propagated ones, hinting at a converging future.

5.3. Clone

A clone consists of plants exclusively produced from a single individual via asexual reproduction. Fruit plants, which are often propagated in this manner, create a group of clones stemming from one original plant through vegetative means. Essentially, all the vegetative descendants of a single plant make up a clone.

5.3.1. Characteristics of Asexually Propagated Crops

- ❏ Perennials dominate (*e.g.*, Sugarcane, fruit trees). Annuals are mainly tubers (*e.g.*, Potato, cassava).
- ❏ Reduced flowering, seed set.
- ❏ Predominantly cross-pollinated.
- ❏ High heterozygosity, inbreeding depression upon selfing.
- ❏ Mostly polyploids (*e.g.*, Sugarcane, Potato, Sweet Potato).
- ❏ Interspecific hybrids exist (*e.g.*, Banana, Sugarcane).
- ❏ Many clones, each variety is a clone.

5.3.2. How Genetic Variation is Obtained in Clones?

- ❏ Genetic diversity within clones can arise from mutation, mechanical blending, and sexual reproduction.
- ❏ Mutation
- ❏ Somatic mutations, referred to as bud mutations, occur relatively rarely. Mutant alleles become homozygous when both alleles within a cell mutate simultaneously or when the mutant allele initiates in a heterozygous state. Dominant mutations are more apparent than recessive ones, requiring homozygosity. Mutations can give rise to chimeras, individuals with multiple genotypes. Techniques are available to generate non-chimera plants from chimeras.

- ❑ **Mechanical mixture**: Mechanical mixture produces genetic variation within a clone, similar to the manner as seen in pure lines.

- ❑ **Sexual reproduction:** Occasional sexual reproduction leads to segregation and recombination. The seedlings obtained from sexual reproduction are genotypically different from the asexual progeny.

5.4. Breeding Methods for Improvement of Clonal Crops

Clonally propagated crops, unlike sexually propagated ones, rely on asexual reproduction, offering distinct advantages. A single outstanding plant can give rise to a new variety, with genotype being less significant due to the absence of sexual reproduction. These plants can originate from an existing variety, enhanced through variability, or created through crossbreeding. This is the essence of clonal crop breeding.

- ❑ Creating genetic variation (natural or hybridization)
- ❑ Selecting best genotype for superior clone or variety.

5.4.1. Clonal Selection

A plant's value comes from genotype (G), environment (E), and their interaction (GE). Only G is heritable. Picking quantitative traits from single observations isn't dependable. For yield and polygenic traits, unreplicated clonal plots mislead. Trusted clone evaluation needs replicated yield trials. Yet, highly heritable traits like height, flowering time, color, disease resistance can be chosen effectively from individual plants or plots.

5.4.2. Hybridization Between Clones

Clonal crops, cross-pollinated, may have self-incompatibility. Chosen parents used for single/multi-parent crosses. For selecting among F_1: when parent value and combining ability are unknown, multiple crosses ensure good F_1. Holds for less improved species. A subset of F_1 grown, potential noted. Inferior discarded. Promising families with exceptional individuals grown larger for selection. Saves time, space, labour by evaluating small populations of crosses initially. Selection within F_1 like clonal selection.

5.5. Genome-wide Selection/ Genomic Selection

Genomic Selection (GS) is a specialized version of Marker-Assisted Selection (MAS). It relies on genotype data from markers spanning the genome to guide selection. Effects of all markers across the genome are assessed, whether significant or not. These marker effects help compute Genomic Estimated Breeding Values (GEBVs) for individuals/lines, forming the basis of selection. An individual's GEBV is the sum of effects from its marker alleles, as per the applied GS model. Breeding Value (BV) predicts progeny phenotype and is based on additive genetic effects via progeny testing. In contrast, Genotypic Value predicts phenotype from genotype, considering both additive and nonadditive genetic effects.

Since the 1980s, phenotype data and kin information have computed estimated breeding values (EBVs) for selection in animal and plant breeding, recently incorporating marker data. Merging markers linked to known QTLs with phenotypes significantly enhanced gains in animal breeding. Incorporating QTLs in GS amassed them more using gene-assisted genomic selection, surpassing standard ridge regression in frequency.

Genomic selection is practiced in different crops like potatoes (*Solanum tuberosum*), cassava (*Manihot esculenta*), sweet potatoes (*Ipomoea batatas*), sugarcane (*Saccharum officinarum*), olive (*Olea europaea*), strawberries (*Fragaria × ananassa*), raspberries (*Rubus idaeus*), grapes (*Vitis vinifera*) and apples (*Malus × domestica*).

5.5.1. Framework of GS

Genomic Selection (GS) employs genome-wide markers to estimate effects, creating a model. This model computes Genomic Estimated Breeding Values (GEBVs) for individuals, aiding selection of superior ones. GS uses two populations: a Training Population (TP) with genotype and phenotype data, and a Breeding Population (BP) with only genotype data. TP's data shapes a predictive model for GEBVs and estimates marker effects in BP. Similarity of alleles linked to target traits in TP affects GEBV accuracy for BP. High GEBV-ranked individuals with favourable alleles become candidates for breeding high-performance varieties (Desta and Ortiz 2014; Xu *et al.* 2020).

5.5.2. A Generalized Procedure for Genomic Selection

Genomic Selection (GS) employs two linked populations: training and breeding. The training population trains the GS model and calculates marker effects for Genomic Estimated Breeding Values (GEBVs) in the breeding population. The breeding population undergoes GS for desired enhancement, isolating superior lines for new varieties or parents.

- ❏ Create suitable training population for the breeding one.
- ❏ Genotype training individuals/lines with numerous evenly distributed genome markers.
- ❏ Thoroughly phenotype trainees for target traits across replicated trials and locations.
- ❏ Use phenotype and marker data for model training, retaining parameter estimates for future use.
- ❏ Evaluate breeding population using same markers without phenotyping.
- ❏ Calculate GEBVs for breeding population from genotypes and training population's marker effects.
- ❏ Select superior individuals/lines based on GEBV estimates.

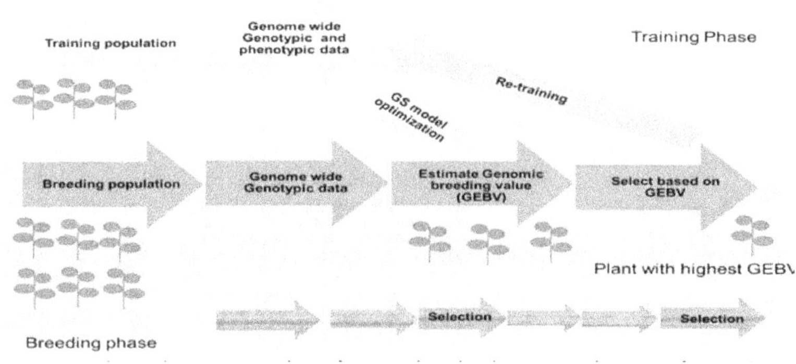

Figure 1. Schematic representation of Genomic selection (Meuwissen et al., 2001)

5.5.3. GS and its advantages over traditional methods and MAS

Introduced in 2001 for cattle breeding, Genomic Selection (GS) has been successfully extended to accelerate breeding in crops and trees. GS represents an advanced evolution of Marker-Assisted Selection (MAS). Unlike MAS, which relies on a subset of markers, GS employs genome-wide markers to estimate genetic values (GEBVs) for untested populations. While MAS requires gene/ QTL mapping and may miss subtle effects, GS doesn't necessitate this mapping and can identify all QTLs, even minor ones. GS captures genetic variations using markers in linkage disequilibrium with QTLs, and its accuracy hinges on a strong linkage disequilibrium between markers and QTLs.

The advantages of GS have been evident in various crops. In maize and wheat, GS has delivered significant genetic gains compared to MAS. For instance, maize grain yield improved by approximately 20% under drought stress conditions. Buckwheat's selection index increased by 21% with GS, surpassing the 15% improvement achieved with phenotypic selection (PS). Wheat's grain yield and agronomic traits advanced by 10% through GS. The cost-effective genotyping made possible by SNP chips and Next-Generation Sequencing (NGS) has enhanced the feasibility of GS. Escalating resource costs have limited large-scale field testing, making GS a more efficient, environmentally friendly, and resource-saving alternative to both MAS and PS.

5.5.4. Genomic Selection in Clonally Propagated Crops

Clonal selection (CS) models, frequently employed in traditional breeding practices, primarily concentrate on selecting traits based on their observable characteristics (phenotypes) and do not take into consideration the existence of both additive and nonadditive genetic influences in crops that are propagated through cloning. These models do not incorporate the distinctive genetic attributes inherent to clonal crops, including heterozygosity, the presence of multiple alleles at a given genetic locus, and the inheritance of stable genetic traits through cloned offspring.

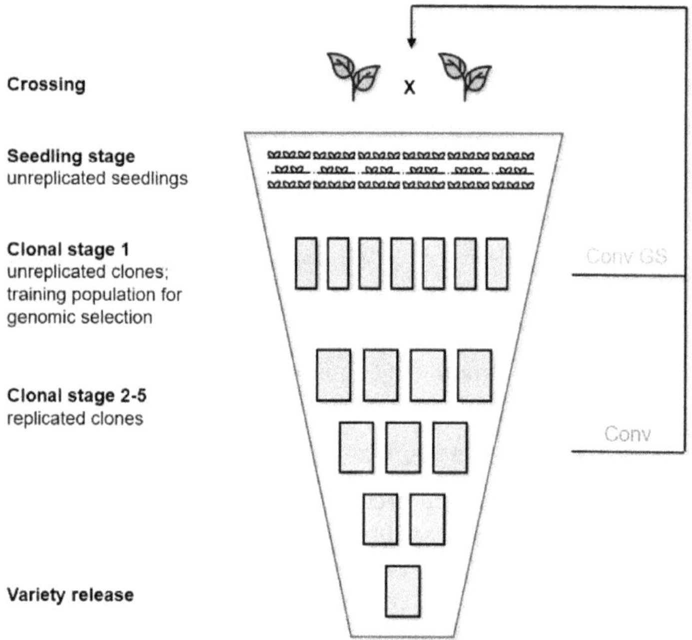

Hence, CS models are not suitable for effectively enhancing clonal crops, and alternative approaches such as genomic selection (GS) should be considered. Given the growing concerns regarding food production, genomic selection holds the potential to reduce the time required for breeding cycles, enhance selection efficiency, and enable more effective utilization of genetic diversity to accelerate genetic improvements in breeding programs. Genomic selection simultaneously assesses numerous genetic loci, haplotypes, or marker effects across the entire genome to calculate genomic estimated breeding values (GEBVs). These GEBVs can be employed to expedite recurrent selection strategies. Genomic selection offers several advantages and surmounts key challenges associated with marker-assisted selection (MAS) based on quantitative trait loci (QTLs). The accuracy of GEBVs relies on factors such as genotype data quality, phenotype data quality, environmental covariates, trait heritability, population structure, model accuracy, and the relatedness between the training and breeding populations. Various statistical models, each making different assumptions about marker effects and their relationships, are employed in genomic selection. The choice of model depends on the traits being considered and the genetic structure of the crops. Selecting the appropriate model for each trait is crucial for achieving accurate predictions.

For quantitative traits with assumed additive effects, mixed models like Genomic Best Linear Unbiased Prediction (G-BLUP) and Ridge Regression

Best Linear Unbiased Prediction (RR-BLUP) are used. On the other hand, for traits influenced by a few major genes and exhibiting both quantitative and qualitative characteristics, Bayesian models such as BayesA, BayesB, BayesCπ, and Bayesian LASSO are suitable. Conventional breeding for self and cross-pollinated crops relies on homozygous parents, while clonal crop breeding, like in strawberries, utilizes heterozygous parents to generate varieties or hybrids. Genomic selection models are not applicable to clonal crop improvement because these crops have heterozygous genotypes with both additive and non-additive gene effects, particularly dominance, which is significant in clonally propagated crops. To effectively enhance clonally propagated crops, genomic selection strategies should incorporate both additive and dominance values. Models like reproducing kernel Hilbert space (RKHS) and random forest (RF) can be applied to efficiently utilize both additive and dominance variance simultaneously. An important benefit of genomic selection (GS) is its capacity to simultaneously enhance multiple traits using a selection index, similar to what breeders employ. GS efficiently evaluates all markers across a population, even those with minor effects, provided there is comprehensive genome-wide marker coverage. Unlike traditional marker-assisted selection (MAS), GS doesn't rely on prior knowledge of the positions of quantitative trait loci (QTLs) on linkage maps. GS employs a marker panel with sufficient density to ensure that all QTLs are expected to be in linkage disequilibrium (LD) with at least one marker locus. The major advantage of GS over MAS is that it can potentially capture all the genetic variance, as marker effects don't need to surpass a significance threshold to be used for predicting breeding values.

5.5.5. Genomic selection is utilized in improvement of clonal crops under two scenarios

Selection of parents: In traditional breeding, selection primarily relies on phenotypic performance, and the choice of parents for the next generation occurs later in the breeding program, resulting in extended generation intervals. Genomic selection (GS) can be highly effective in clonally propagated crops with diploid-like meiotic behavior when selecting parents for crossing is based on genomic predicted cross-performance (GPCP), assuming dominance plays a minor role. However, in clonal breeding programs, selecting parents based on GPCP becomes essential to accelerate genetic improvements and mitigate inbreeding, especially as dominance becomes more significant. An innovative two-phase breeding approach that incorporates parent selection based on GPCP has been instrumental in advancing clonally propagated crop breeding (Werner et al., 2023). GS models like G-BLUP, RR-BLUP, BayesA, BayesB, BayesCπ, and Bayesian LASSO can effectively aid in the selection of new parents for the crossing stage, particularly when additive genetic variation is crucial.

Identification of best genotype: Genomic selection has significant potential for enhancing the selection of top-performing clones in variety development. It leverages correlations between genomic markers and traits to predict genotype value based on marker profiles. To effectively apply genomic selection in these crops for variety development, methods such as RKHS and RF, capable of modeling both additive and nonadditive effects, would be necessary.

5.5.6. Advantage of Using Genomic Selection Over Conventional Breeding in Clonal Crops

In traditional breeding, genetic progress eventually plateaus, making it challenging to develop superior germplasm unless driven by significant factors like new diseases or market demand for specific traits. Genomic selection (GS) offers a major advantage by drastically shortening breeding cycles compared to traditional methods, reducing the cost of extensive phenotyping, and thereby accelerating genetic gains, ensuring food and nutritional security (Heffner et al., 2010). However, GS's success depends on factors such as the size and diversity of training and breeding populations, trait heritability, genotype-environment interactions, marker density, and genetic relationships, all of which affect the accuracy of genomic predictions and consequently genetic gains. Genetic gain in a breeding program is determined by selection intensity and accuracy, trait heritability, trait variance, and the generation interval. GS primarily focuses on reducing time, increasing selection accuracy and intensity to enhance genetic gains.

The breeder's favorite equation

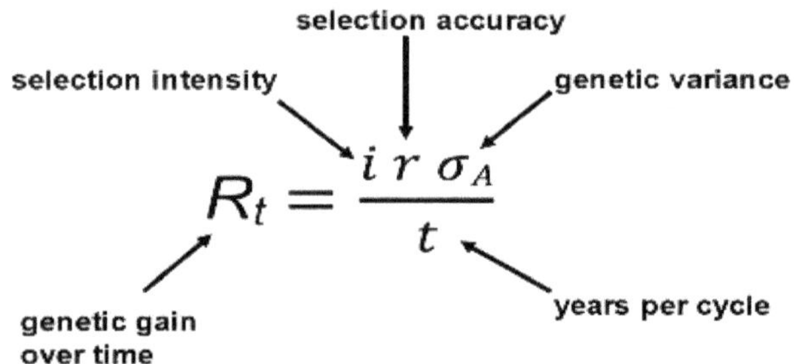

- ☐ **Reduction of generation interval:** By selecting parents immediately upon genotyping, genomic selection reduces the generation interval, accelerates breeding, and allows for quicker recycling of individuals based on genomic estimated breeding values (GEBV), which are more effective than phenotype-based selection. Implementing GS has the potential to achieve significant progress, enabling multiple selection rounds within conventional single-round timeframes, especially when accurate GEBVs can be predicted for crucial traits like yield and quality. However, in clonally propagated species, GS could potentially reduce the generation interval to one year or less by selecting parents right after genotyping, though several challenges, including DNA marker resources and polyploid genomes, need to be addressed. As genotyping-by-sequencing (GBS) costs decrease and improved methods for genotyping polyploid genomes emerge, barriers to marker availability are expected to diminish. Applying genomic selection at the seedling stage could yield up to five times the genetic gain achieved in conventional strawberry breeding, provided that selection intensity, diversity, and selection accuracy remain constant (Grüneberg et al., 2009), thanks to reduced generation intervals and enhanced selection accuracy when choosing seedlings based on their predicted performance as clones rather than relying solely on their phenotypic traits.

❑ **Increased selection accuracy:** In early breeding stages with limited locations and replications, genomic estimated breeding values (GEBV) can significantly enhance selection accuracy compared to traditional phenotype and pedigree-based methods. However, when applying genomic selection (GS) in clonally propagated crops, the presence of dominance can impact breeding programs. Clonally propagated crop genotypes are typically heterozygous, and their genetic value results from both additive and non-additive gene effects. Assuming negligible epistasis, the non-additive gene action is primarily represented by dominance variance. While long-term genetic gains are primarily defined by additive genetic components, it is crucial in clonally propagated crops to increase additive value over time while simultaneously preserving dominance value through parental selection and recombination.

❑ **Selection intensity:** Selection intensity can be elevated, such as by genotyping and predicting a greater number of genotypes than can be practically tested in the field. Consequently, breeders have the capacity to select more individuals based on their genetic profiles than they can assess through physical phenotype evaluation. This is facilitated by the cost-effectiveness of genotyping compared to phenotyping, allowing for the evaluation of a larger pool of potential candidates for selection.

5.5.7. Application of Genomic Selection in Clonal Crops

Highly informative SNP marker panels in yam (*dioscorea rotundata*) for genomic selection and prediction for genomic prediction and selection. This set of selected markers can also be used for the Genomic Prediction of Cross Performance (GPCP) (Agre *et al.*). Although the heterozygous nature of potato, completion of the potato genome sequence and the identification of a large number of genome-wide SNPs paved a way for application of genomic selection in potato. Rodriguez *et al.* (2018) found that an additive model could account for a sizable (>90%) of the total genetic variance for both late blight and common scab in potato which had many genomic selection studies. small (but statistically significant) gains in prediction accuracy when accounting for dominance with small (but statistically significant) gains in prediction accuracy when accounting for dominance. Genomic predictions were calculated using the BayesA and BayesB models. Three African cassava breeding institutions, namely NaCRRI, NRCRI, and IITA, have implemented genomic selection (GS) in their programs and analyzed data from their respective GS programs using different germplasm collections. The study conducted by Wolfe *et al.* (2016) provide the details on the origins and pedigrees of the NaCRRI and NRCRI clones used in the analysis. This study also highlights the importance of factors such as trait heritability, prediction models, and training population composition in determining the accuracy and rate of genetic gain progress in genomic selection. Yonis *et al.* (2020) assessed the viability of the picture phenotyping technique as well as GWAS and genomic prediction for image-extracted size and shape attributes. Root size and form image-extracted trait prediction accuracies were often greater than yield trait prediction accuracies. The study suggests that at least a few big impact loci influence root-related phenotypes and that introgressed genome areas include QTL for root yield, size, and shape features. Habyarimana *et al.* (2014) evaluated prediction accuracy for yield, yield components, and quality attributes in potato using three GS models: Bayesian LASSO, genomic best linear unbiased prediction (G-BLUP), and reproducing kernel Hilbert space (RKHS). They showed r > 0.60 prediction accuracy for numerous parameters such as carotenoids, tuber dry matter, and overall yield.

5.6. Challenges of Conventional Breeding in Clonal Crops:

Clonal crops, which are propagated asexually through vegetative parts like cuttings or tubers, present unique challenges in conventional breeding compared to sexually reproducing plants. Some of these challenges include

- ❐ **Limited Genetic Diversity:** Clonal crops have limited genetic diversity since they are propagated from a single-parent plant. This can make them more susceptible to diseases, pests, and environmental changes.

- ❐ **Difficulty in Introducing New Traits:** Traditional breeding relies on crossing plants with desired traits to generate genetic variation. In clonal crops, introducing new traits can be challenging as there is no sexual recombination. Genetic modification techniques might be needed to incorporate specific traits.

- ❐ **Disease Vulnerability:** Clonal propagation can lead to the rapid spread of diseases through the entire crop population. If a disease affects one plant, it can easily affect all plants since they are genetically identical.

- ❐ **Economic and Time Investment:** Developing new clonal varieties can be time-consuming and resource-intensive. Each new variety requires extensive testing and validation before it can be released to farmers.

- ❐ **Regeneration and Mutations:** Over time, clonal crops can accumulate mutations, potentially leading to undesirable changes in traits. Regeneration of plants from tissue culture can also result in somaclonal variation, where plants differ from the original clone.

❐ **Environmental Adaptation:** Clonal crops might have limited ability to adapt to changing environments due to their genetic uniformity. This can be a concern as climate change and other factors alter growing conditions.

5.7. Challenges of Genomic Selection in Clonal Crops

5.7.1. Genetic Diversity and Linkage Disequilibrium

The challenge of limited genetic diversity due to asexual reproduction in clonal crops can significantly impact the effectiveness of genomic selection. Genetic diversity is crucial for the accuracy of genomic predictions, as it allows for the identification of markers associated with important traits. However, clonal reproduction restricts the introduction of new genetic variations, resulting in reduced levels of linkage disequilibrium (LD), which is the non-random association of alleles at different loci. Low LD hinders the ability to precisely pinpoint trait-related markers, making it more challenging to accurately predict the performance of clonal plants using genomic information. This challenge underscores the need for innovative strategies to overcome the limitations imposed by clonality and enhance the applicability of genomic selection in clonal crops.

5.7.2. Marker Density and Coverage

Ensuring accurate genomic predictions relies on sufficient marker coverage throughout the genome. Yet, in clonal crops, selecting suitable marker density is complex due to unique genetics and lower recombination rates. Inadequate markers can lead to less reliable predictions. This challenge arises from distinct genetic traits and reduced recombination. Comprehensive genetic markers spanning the genome are vital for precise predictions. However, clonal crops, reproducing asexually, have limited recombination compared to sexually reproducing plants. This hampers genetic material shuffling and leads to regions with higher linkage disequilibrium (LD). Selecting the right marker density becomes challenging, as it must capture unique genetic diversity. Inadequate markers in this unique genomic setting can compromise prediction accuracy. To address this, researchers must strategize marker selection, focusing on higher recombination regions or advanced sequencing for improved marker resolution in clonal crops.

5.7.3. Genotype by Environment Interaction

Clonal crops might exhibit significant genotype by environmental interactions, where the performance of a genotype is highly dependent on specific environmental conditions. Genomic selection models need to account for these interactions, requiring extensive and accurate environmental data, which might be challenging to obtain. Clonal plants are often adapted to specific growth conditions, making their performance-sensitive to changes in factors like temperature, soil quality, and humidity. When applying genomic selection to these crops, accurate predictions of a genotype's performance across different environments become essential. However, gathering comprehensive and precise environmental data is challenging, as it requires extensive monitoring and recording of various parameters. Inaccurate or incomplete environmental information could lead to biased predictions and hinder the success of genomic selection efforts in clonal crops. Developing strategies to collect and integrate diverse environmental data is crucial to enhance the robustness and applicability of genomic selection models in these crops.

5.7.4. Validation and Training Populations:

Developing structured training and validation populations is crucial for successful genomic selection, especially for clonal crops with their distinct genetic traits. However, this task is intricate due to the unique nature of clonal crops. As these crops reproduce asexually, their genetic diversity is limited and can be traced back to a few ancestors. Creating diverse and representative populations that encompass the genetic variation within clonal crops is a challenge. Without a well-designed approach, there's a risk of overfitting, where models perform well on training data but fail to generalize, and biased predictions that don't reflect real-world complexity. Striking the right balance between capturing limited genetic diversity and avoiding over fitting or biases is vital. This requires selecting representative individuals, sampling strategically from diverse environments, and using statistical techniques for clonal populations. By addressing these challenges, researchers can enhance genomic selection accuracy for clonal crops, improving trait enhancement efforts.

5.7.5. Phenotyping Challenges

Population structure in clonal crops can be intricate due to factors like historical clone introductions and selection events. These events can result in diverse genetic backgrounds within the population. Managing and modeling this complexity is crucial to prevent misleading results in genomic selection analyses. If not properly accounted for, population structure can introduce confounding effects, leading to erroneous associations between genetic markers and traits. To address this challenge, researchers employ statistical techniques that correct for population structure, ensuring that the observed associations are genuine and not artifacts of genetic diversity. By accurately modeling population structure, genomic selection analyses can yield more reliable predictions and facilitate the identification of trait-associated markers in clonal crops.

5.7.6. Population Structure

Population structure in clonal crops can be intricate due to factors like historical clone introductions and selection events. These events can result in diverse genetic backgrounds within the population. Managing and modeling this complexity is crucial to prevent misleading results in genomic selection analyses. If not properly accounted for, population structure can introduce confounding effects, leading to erroneous associations between genetic markers and traits. To address this challenge, researchers employ statistical techniques that correct for population structure, ensuring that the observed associations are genuine and not artifacts of genetic diversity. By accurately modeling population structure, genomic selection analyses can yield more reliable predictions and facilitate the identification of trait-associated markers in clonal crops.

5.7.7. Handling Rare Alleles and Mutations

Addressing rare alleles and mutations in clonal crops presents unique challenges due to their potential influence on valuable traits, despite their limited occurrence within the population. Detecting and integrating these variants into genomic selection models is complex, as standard approaches may not effectively capture their significance, leading to their underrepresentation in predictive models. Specialized strategies are essential, such as targeted genotyping and custom statistical methods prioritizing the impact of rare alleles. These approaches enable the identification and utilization of precious genetic resources in clonal crops, ultimately enhancing the accuracy of genomic selection for trait improvement. In conclusion, conventional breeding techniques offer both opportunities and challenges in clonal crops. While exploiting the vast genetic resources and adapting traditional breeding methods are advantageous, the inherent limited genetic diversity and extended breeding cycles pose obstacles. Innovative methods like mutation breeding and genomics-assisted breeding can complement conventional approaches, fostering sustainable enhancement in clonal crops and ensuring agricultural sustainability and food security.

5.8. Conclusion

In conclusion, the challenges posed by clonality in the conventional breeding of crops are substantial and multifaceted. The inherent lack of genetic diversity resulting from asexual reproduction restricts the pool of available genetic variations, limiting the potential for creating novel traits and enhancing adaptability to changing environments. The accumulation of mutations over generations, susceptibility to diseases and pests, labor-intensive propagation methods, and genetic instability further compound the difficulties associated with conventional breeding in clonal crops. Addressing these challenges demands innovative solutions, such as harnessing advanced breeding techniques and technologies to introduce genetic diversity and enhance the genetic stability of clonal crops. By acknowledging and actively tackling these challenges, researchers and breeders can pave the way for more resilient and productive clonal crop varieties, ensuring food security and sustainable agriculture in the face of evolving global demands.

References

Agre, P.A. and Asfaw, A., Highly Informative SNP Marker Panels in Yam (Dioscorea rotundata) for Genomic Selection and Prediction. https://doi.org/10.25502/xjtv-mq33/p

Bradshaw, J. E. and Bradshaw, J. E., 2016. Genetic structure of landraces. *Plant breeding: past, present and future*, 9(11), 273-290. https://doi.org/10.1007/s10681-016-1815-y

Grüneberg, W., Mwanga, R., Andrade, M. and Espinoza, J., 2009. Selection methods. Part 5: Breeding clonally propagated crops. *Plant breeding and farmer participation*, 275-322.

Habyarimana, E., Parisi, B., Onofri, C., Govoni, F. and Mandolino, G., 2014. Efficiency of genomic selection for yield, marketing and nutritional quality traits in hundred thirty-nine cultivated potato genotypes. In EAPR 2014 Brussels-19th Triennial Conference of the European Association for Potato Research (55-56).

Meuwissen, T., 2003. Genomic selection: the future of marker assisted selection and animal breeding. *Marker Assisted selection: a fast track to increase genetic gain in plants and animal breeding*, (54-59).

Rodriguez, F., Douches, D., Lopez-Cruz, M., Coombs, J. and de Los Campos, G., 2018. Genomic selection for late blight and common scab resistance in tetraploid potato (Solanum *tuberosum*). *G3: Genes, Genomes, Genetics*, 8(7), 2471-2481. https://doi.org/10.1534/g3.118.200273

Singh B. D. Plant Breeding: Principles and methods. 12, Kalyani publishers, New Delhi, 2021, 359.

Werner, C.R., Gaynor, R.C., Sargent, D.J., Lillo, A., Gorjanc, G. and Hickey, J.M., 2023. Genomic selection strategies for clonally propagated crops. *Theoretical and Applied Genetics*, 136(4),74. https://doi.org/10.1007/s00122-023-04300-6

Wolfe, M.D., Rabbi, I.Y., Egesi, C., Hamblin, M., Kawuki, R., Kulakow, P., Lozano, R., Carpio, D.P.D., Ramu, P. and Jannink, J.L., 2016. Genome-wide association and prediction reveals genetic architecture of cassava mosaic disease resistance and prospects for rapid genetic improvement. The *Plant Genome*, 9(2), 2015-11. https://doi.org/10.3835/plantgenome2015.11.0118

Yonis, B.O., Pino del Carpio, D., Wolfe, M., Jannink, J.L., Kulakow, P. and Rabbi, I., 2020. Improving root characterisation for genomic prediction in cassava. *Scientific Reports*, 10(1), 8003. https://doi.org/10.1038/s41598-020-64963-9 .

Chapter-6

Population Genetics and Plant Breeding

M. Hemalatha[1*], B. Kiranmayee[2]

1. *M.Sc. in Genetics and Plant Breeding, IGKV, Raipur*

2. *Ph. D. Scholar in Genetics and Plant Breeding, IGKV, Raipur*

6.1. Introduction

Population genetics falls under the umbrella of genetics that plays a fundamental role in gaining insight into the genetic underpinnings of evolution. It examines genetic variations within and between populations, shedding light on how species evolve and adapt over time through changes in allele frequencies within specific groups of organisms.

Plant population genetics helps us study the genetic disparities that are located internally and between plant populations. These differences can be crucial for a species' adaptation and survival in different environments. It focuses on quantifying the manner in which an allele's frequencies operate (different versions of genes) change over generations within a particular population. This information is vital for understanding how populations evolve. The arena of population genetics, which heavily relies on mathematical models, emerged in the nascent to mid-20th century. Much of the cornerstone of the research was laid before the 1980s, although the discipline has unabatedly to evolve with advancements in genetic research. Mendelian principles of inheritance, such as segregation and random mating, form the foundational framework of population genetics. These principles help explain how genetic diversity is maintained within populations. Conservation genetics is a specialized area that applies population genetic theory to measure and preserve genetic diversity within target populations of plants (and animals). It helps conservationists make informed decisions about preserving endangered species and maintaining healthy ecosystems. Population–level plant genetics is a vital field that enables scientists to explore the genetic heterogeneity in plant groups, understand how this variation contributes to adaptation and evolution, and use this knowledge for conservation efforts to protect plant species and their genetic diversity.Top of Form

6.2. Population and Gene Pool

A Mendelian population encompasses individuals characterized by equivalent species sharing a mutually possessed genes and living amidst a geographic area sufficiently restricted so that any member mirrors the alike potential to mate with another member (of opposite sex) pertaining to the similar population, i.e. random mating. On the flip side, a gene pool encompasses all the genes and variations within a population that engages in sexual reproduction, which can be passed down for the subsequent cohort. Instead of focusing on specific peculiarities, population-level genetics examines the prevalence rates of different with the passage of time; the genetic content of this gene pool undergoes alterations. Cognizance the structure of populations is crucial for both traditional and unconventional breeding methods.

It's worth noting that with genetic engineering, gene transfer across different biological boundaries becomes possible. In instances where cross-pollinated species, breeding efforts primarily aim to enhance communities as a whole rather than individual plants, which differs from self-pollinated species where the focus is often on improving individual plants. To grasp the significance of population structure in the discipline of plant breeding, it's essential to comprehend the manner of distinction present, the genetic mechanisms behind it, in conjunction with the tools utilized to modify the genetic makeup.

6.2.1. Mathematical Conceptualization of a Gene Pool

As mentioned earlier, gene prevalence is a fundamental notion in genetic analysis of populations. It deals in conjunction with both genetic makeup among a population and how genetic material is passed on to the thereafter generation. In order to delineate the genetic constitution of a population, we use a set of gene frequencies. Numerous critical elements impact the transmission of genes from one era to the next within a population. These factors include the population magnitude, differences in the ability to reproduce and survive, migration events and genetic mutations, furthermore way individuals choose their mates.

Genetic the occurrence frequencies might fluctuate across generation unto the subsequent due to natural fluctuations. Plant breeders play a crucial role in guiding the metamorphosis of breeding populations. They do this by selecting specific parents to start a breeding program, determining how these parents are mated, and employing artificial selection to choose which individuals will contribute in the upcoming generation.

It's important to note that while the genetic profile of a population persists over the course of generations, the specific combinations of these genes (genotypes) change with each new generation. Plant breeders routinely engage with populations where the bestowal of attributes doesn't neatly follow Mendelian segregation patterns, even though, in reality, they still adhere to Mendelian principles. Mendel's work primarily dealt with genes that had clear-cut effects and were easily categorized in offspring ratios.

In contrast, plant breeders often deal with populations where differences are measured on a spectrum rather than distinct categories. To understand and describe these population phenomena, population genetics relies on mathematical models. However, creating these models requires making conjectures concerning the population and its environment.

In summary, population genetics involves studying how genes are passed on in populations, and mathematical models help us make sense of these complex dynamics by outlining specific notions about how populations and their environments interact.

To calculate gene frequency, let's consider a substantial population engaging with haphazard mating occurs, and there are no mutations, gene flow to or from other populations, no specific advantage for any gene type, and normal meiosis. We're looking at one gene location, let's call it "A," with two forms, "A" and "a." In this gene pool, the proportions of "A" is represented as "p," while the rate of "a" is "q." It's imperative to underscore that the sum of these frequencies, "p" and "q," adds up to 1, which represents the complete genetic reservoir.

Imagine a population of "N" pairs of chromosomes (diploids) in which these two gene forms, "A" and "a," exist at one specific location. Assuming that one of the aforementioned gene forms, let's say "A," is dominant, we can have three possible genetic composition within the offspring: "AA," "Aa," and "aa."

Now, let's denote the incidences of these genotypes as "D" for "AA," "H" for "Aa," and "Q" for "aa." Since each individual in this population has two gene copies, we can calculate the cumulative count of "A" gene forms in the entire population as 2D+H. The share or prevalence of A alleles (tagged as p) in the population is obtained as follows:

$$2D + H/2N = (D + 1/2H)/N = p$$

In simpler terms, we find the percentage or prevalence of the "A" alleles, which we represent as "p," in the population using this formula. Allele a can undergo the same process, with the symbol q.

$$\text{Further, } p + q = 1 \text{ and ergo } p = 1 - q.$$

6.3. Hardy–Weinberg Equilibrium

Contemplate a population engaging in random mating, where each male gamete possesses an equal probability of mating with any female.

When individuals with different gene variations (in this case, "A" and "a") randomly mate, their offspring can have three possible gene combinations: "AA," "Aa," and "aa." Each of these combinations has a specific manifestation in the population: "p^2" for "AA," "2pq" for "Aa," and "q^2" for "aa."

It's crucial to emphasize that sum pertaining to these gene frequencies should always equal 1, meaning they cover the entire population. To put it another way, "$p^2 + 2pq + q^2$" must equal 1. This mathematical relationship is known as the Hardy-Weinberg equilibrium.

Hardy and Weinberg, scientists from England and Germany, discovered that this equilibrium occurs amidst extensive populations. They found that the rates of different gene combinations are contingent on in a population on the gene prevalence in the previous era, omitting the frequencies pertaining to a precise gene combination themselves. This principle helps us understand how gene variations are maintained within populations over time.

The Hardy-Weinberg equilibrium equation is typically written as:

$$p^2 + 2pq + q^2 = 1 \text{ (or 100\%)}$$

The Hardy-Weinberg equilibrium remains unwavering when certain requirements met:

- ❏ Random mating among a vast diploid population.
- ❏ Allele A alongside allele a have equal fitness; one doesn't provide a competitive advantage over the other.
- ❏ No differential migration occurrence of alleles within or beyond the population.
- ❏ Mutation rates for allele A and allele a are the same.

In essence, these conditions mean amidst the population with random mating, genetic variability remains consistent through the passage of generations next. The zenith occurrence of heterozygotes (Aa) cannot exceed 0.5.

The Hardy-Weinberg law proclaims that equilibrium comes about as achieved through any gene locus after just a generation characterized by random mating. Within the scope plant breeding perspective, two modes are apparent in variability: a homozygous pair (AA and aa), known as **"free variability,"** which can be selectively fixed, and the intervening heterozygote (Aa), called **"hidden or potential variability,"** which can yield novel variation through the mechanism of segregation.

In the realm of outcrossing species, homozygotes can cross to produce more heterozygotic variation. With random unbiased mating with no selective criteria, the rates pertaining to mating and the segregation balance out to maintain a 50%:50% ratio of available and potential variability. This means the layout of the population is dynamic, continuously undergoing segregation and mating.

Nevertheless, in the event that 2 loci (gene locations) are involved, equilibrium is achieved more slowly, especially when there's strong genetic linkage between the loci.

In most plant traits that affect advancement, reproduction and growth, the variation we see is quantitative. This means it's steered by a broad spectrum of genes, not just 1 or a handful. These groups of genes, called polygenes, behave similarly to regular Mendelian genes in terms of preeminence, modifier gene effect (epistasis), and how they're linked together on chromosomes. We can use the Hardy-Weinberg equilibrium to understand these traits, but it's a bit more complicated to demonstrate.

Now, let's talk about a different type of variation. Imagine you have two separate gene locations, each with two forms, let's call them "A" and "a" for one and "B" and "b" for the other. Assuming there's no dominance or gene interactions, we can end up with nine possible combinations of these genetic repertoire within the population when individuals mate randomly. These combinations result in different genotypes and appearances in a certain frequency pattern.

In simpler terms, these gene combinations can lead to different traits in plants, and their frequencies follow a specific pattern when plants mate randomly. Some of these combinations are a source of readily visible variation, while others contain hidden potential for variation. This hidden potential only becomes visible when specific conditions, like crossbreeding and the next generation's segregation, occur.

This hidden potential is a bit tricky. Unlike when we're dealing with just one gene location, where half the variation is at your disposal for selection in next generation, here, only one-eighth of the variation can be selected directly. The rest remains veiled within the genetic makeup of the plants, either in heterozygous or homozygous forms.

Now, let's mull over the scenario when you have more genes involved, and they're linked unified in a specific way. The flow of variation between homozygous potential (hidden) and free (visible) states depends on how closely these genes are linked. If they're tightly linked, the variation flow is restricted. In plants that mainly self-pollinate or in breeding where inbreeding is encouraged, the proportion of heterozygotes decreases significantly in each generation, nearly disappearing over time.

However, in plants that cross-pollinate, each plant in the population can have both hidden and visible genetic variation. Plant breeders make use of this genetic diversity to improve plant populations. They also consider factors like adaptability, fitness, and other desirable traits when selecting plants for breeding programs. The goal is to shift the population's genetic structure towards meritorious traits for the plant and for human needs. The Hardy-Weinberg equilibrium entails certain prerequisites that need to be met. Notwithstanding, there are cases where these conditions are approximately satisfied.

6.3.1. Population Size

The Hardy-Weinberg equilibrium typically mandates a significant population where individuals randomly mate to hold true. This condition is generally met customarily found in cross-pollinated species, barring instances of non-random mating behaviors, like inbreeding or assortative mating, come into play. Inbreeding naturally occurs in amongst self-breeding species, while assortative mating is empowered to happen when closely spaced cross-pollinated plants interact amidst practitioners.

Effective population size (Ne) is a fundamental parameter in population-level genetics models. It represents the count of breeding individuals in an idealized population that exhibits the same allele frequency experiences shifts through random genetic drift or inbreeding as the actual population under consideration. Ne can be seen as the breeding population size, and it's distinct from the minimum viable population size (MVP), which estimates the smallest number of individuals needed to sustain a population over an extended period without significant intervention.

Determining Ne is challenging due to difficulties in obtaining accurate data for real populations and the complex interplay of aspects shaping the relationship between Ne along with the actual population size (N). However, we can gain insights by examining the impact of various factors.

Let's explore how variation in family size affects the Ne-N relationship. Suppose individuals in a population mate randomly, and each individual contributes a certain number of alleles (equivalent to family size, denoted as ki) to the next generation, with a variance denoted as V. Differences in family size arise from chance or variations in the parental micro-environment.

If N holds steady across subsequent era, and each parent contributes an average of two alleles across the span of periods (k = 2), the tie between Ne and N can be approximated as:

$$Ne = (4N - 2) / (2+V).$$

When reproductive outputs differ randomly, ki follows a binomial distribution. The variance (V) can be estimated as:

$$V = 2(1 - 1/N),$$

resulting in Ne approximately equal to N.

In natural populations, family size distribution is often non-random, leading to a variance (V) greater than 2(1 - 1/N), making Ne smaller than N.

In controlled environments like gene banks, where each parent contributes exactly two alleles (V = 0), Ne simplifies to Ne = 2N - 1.

Different features of interest, including various genetic loci, may result in different Ne values. Ne can also be calculated as within-species genetic diversity divided by four times the mutation rate, as heterozygosity in such an idealized population equals 4Nμ.

In summary, Ne is a critical concept in population genetics, representing the breeding population size that mirrors the genetic dynamics of the actual population. It's often smaller than the absolute population size (N) due to aspects including non-random family size.

6.4. Multiple Loci

Research has unveiled the prospect that alleles at 2 can different gene locations to have random mating proportions deviating from harmony with each other. Furthermore, evenness between two gene locations isn't achieved after just one cycle of random mating, like the Hardy-Weinberg law originally suggested. Instead, it happens morphing slowly across numerous eras. Additionally, the disclosure of genetic linkage between these gene locations can prolong the deceleration of reaching equilibrium.

To exemplify, in case there's no genetic linkage (c = 0.5), discrepancy in the real frequency and equilibrium frequency reduces amplify by 50% in every period. At this tempo, it would necessitate approximately 7 eras to approach equilibrium. However, with c = 0.01 and c = 0.001, attainment of equilibrium would require approximately 69 and 693 generations, respectively. In some cases, a combined gene frequency can be computed for genes positioned on these two gene locations. For instance, given the likelihood at one location (Aa) is 0.2 and at another location (bb) is 0.7, the amalgamated frequency for a genotype with both gene variations (Aabb) would be 0.2 X 0.7 = 0.14.

The magnitude of "c" represents the correlation rate of these 2 loci. Linkage brings about a retardation in the approach speed; the closer the linkage, the more gradual the rate. For "c" equal to 0.5,lack of linkage. In this case, the equilibrium value is approached at a slow pace and is theoretically beyond reach.

6.5. Factors Effecting Gene Frequency

Amendments in the gene frequency of a population could occur influenced by two core varieties of processes:

6.5.1 Systematic and Dispersive

6.5.1.1. Systematic Processes

These lead to predictable fluctuations in gene frequency with respect to both direction and scale. Examples of systematic processes include selection (favoring certain traits), migration (gene flow between populations), and mutation (changes in the genetic code).

6.5.1.2. Dispersive Processes

These operations, often related to inconspicuous populations, result in foreseeable alterations in gene frequency amount but not direction. To rephrase, we can estimate how much the gene frequency will change, but we can't precisely predict which way it will change.

For instance, systematic processes like selective breeding alternatively movement of organisms between different populations can lead to expected shifts in gene frequencies, either increasing or decreasing certain traits. Dispersive processes, put differently, may involve random events that affect gene frequencies without a clear pattern.

6.5.2.1. Migration

Migration, which is significant in small populations, involves individuals coming assimilated into a pre-existing population from an external source. In plants, this typically happens through the transfer of pollen or gametes. The influence of this immigration is contingent on how many new individuals arrive and how different their gene frequencies are juxtaposed with the native population. In simple terms, it can change the genetic makeup amidst the population. Plant breeders employ this procedure when they want to introduce new infuse genes within their breeding populations. Even if the prevalence of the immigrant gene is low, it can still have a significant influence on the host's genetic composition and characteristics they control.

6.5.2.2. Mutation

Natural mutations are not frequently encountered, and a specific mutation usually doesn't have a noteworthy impact on gene frequencies. Mutations often have recessive effects, but sometimes they can show dominant traits. Recurrent mutations, which happen persistently at a steady rate, can affect gene frequencies more significantly. In everyday plant breeding, natural mutations don't play a major role, but breeders might intentionally induce mutations to create new genetic diversity.

Mutations and gene edits can change DNA in various ways, from altering individual DNA letters to rearranging large sections of genes or chromosomes. These changes can be categorized as either non-recurrent (happening only once) or recurrent (occurring at a certain rate). Gene mutations often happen repeatedly, while chromosomal rearrangements are rare and unique events due to random breaks in chromosomes.

Non-recurring mutations may prevail within a population if the individual with the mutation has many descendants. However, the likelihood of these mutations remaining decreases over generations because a single mutation without a selective advantage cannot permanently change the population.

There are mechanisms beyond natural selection that can help non-recurrent mutations increase in frequency in populations. For instance, meiotic drive favors the bequeathal of a 1 allele over another in gametes, and genetic drift can also influence mutation frequency changes.

In a simplified example, let's consider an allele "A" that mutates irreversibly to "a" at a low rate (μ). To ascertain the frequency of A in each generation (p_n), one can compute

$$p_n = p_{n-1}(1 - \mu)$$

Where p_{n-1} is the prevalence in the immediate past generation. When we want to find the equilibrium frequency when the mutation reaches a steady state, it simplifies to:

$$P_n = p_0(1 - \mu)^n$$

Where p_0 is the initial frequency of A and allele A is fixed in n time.

If the mutation is reversible (a to A) with a cadence v, at equilibrium $p\mu = qv$, the equilibrium frequency can be derivable as:

$$p = v \ / \ (\mu + v)$$

Mutation rates are typically low, and most mutations are harmful and get removed pertaining to the population quickly. However, mutation is essential in evolution as it is the ultimate source of genetic variation.

6.5.2.3. Selection

The principal avenue for plant breeders to change gene frequencies in a population is through the process of selection.. It involves choosing plants with specific traits to be parents for the succeeding offspring. This process shifts the average trait values in the succeeding progeny compared to the parental generation.. The degree of alteration is contingent on the trait. under consideration. For instance, breeders might select aiming for greater output while also considering other factors like chemical composition. To enable selection to work well, it is vital for there to be observable differences in the traits among the plants, and these differences should be at least partially due to genetics.

- ❑ **Modes of Selection**: Three significant divisions of selection: stabilizing, disruptive, and directional. The one that matters most to plant breeders is directional selection. These selection types work to different degrees in both artificial and natural selection, but they have different goals. In natural selection, the aim is to improve the overall fitness of a species, while in plant breeding, breeders use artificial selection to achieve specific goals, which may not inevitably entail creating the fittest plants.

- ❑ **Stabilizing Selection:** In nature, selection is an ongoing process. For traits that directly affect a plant's fitness (like its ability to survive and reproduce), selection always moves toward the optimal trait for a given environment. However, for other traits, once the best trait has been reached, selection works to maintain it as long as the environment remains stable. This type of selection is called stabilizing selection. For example, when it comes to flowering time, stabilizing selection favors neither early nor late flowering. This type of selection promotes genetic diversity.

❏ **Disruptive Selection:** Natural habitats are usually not the same everywhere. They have different conditions that lead to different traits being favored. Disruptive selection happens when extreme traits are better for survival and reproduction than the average traits. This type of selection increases diversity. The interconnectedness of these distinct traits and how often genes are exchanged between them determine the genetic architecture of a population. As an illustration, in the human species, the presence of two genders (male and female) exemplifies this type of genetic diversity.

❏ **Directional Selection:** Plant breeders use directional selection to change populations or varieties in a specific way. They focus on particular traits and use techniques like cross-breeding to mix genes from different parents. The goal is to create a novel genetic mix that produces a specific trait. Directional selection often culminates in the establishment of dominance or genetic interaction between genes.

Effect of Mating System on Selection

There are four main mating systems: random mating and non-random mating, which includes genetic assortative mating, phenotypic assortative mating, and disassortative mating.

❏ **Random Mating:** In nature, true random mating is quite rare because selection is almost always happening. However, random mating without selection doesn't Alter gene frequencies but may impact the average trait values within the population.

❏ **Non-Random Mating:** Non-random mating can happen in two main ways: genetic assortative mating, where closely related individuals mate (often seen in self-fertilizing species), and phenotypic assortative mating, where individuals mate based on their physical resemblance in tandem (often leading to two extreme phenotypes).

❏ **Disassortative Mating:** Genetic disassortative mating involves mating Individuals who have a more distant relationship than they would typically have in random mating. Phenotypic disassortative mating happens when individuals with very different appearances are chosen to mate. Either of these strategies is effective for maintain genetic diversity in a population.

These different mating systems play a role in shaping the genetic makeup of a population, and plant breeders choose the one that is most congruent with their breeding goals.

6.5.2.4. Genetic Drift

Genetic drift is a process that causes changes in the frequencies of alleles (gene variants) in a population over time. It's particularly significant in natural populations with a limited number of individuals. Genetic drift occurs due to random sampling and chance events, influencing which individuals survive and reproduce Pivotal to note these shifts are not driven by environmental factors or adaptive pressures; they are prone to be beneficial, neutral, or detrimental to reproductive success. The aftermath of genetic drift are as follows:

- ❑ **Effect of Population Size:** Genetic drift has a more significant impact on smaller populations. In smaller populations, random chance can lead to rapid shifts in allele frequencies. In contrast, larger populations are more stable concerning frequencies of alleles.

- ❑ **Genetic Diversity:** Genetic drift is permitted to lead to the loss of alleles from a population. Therefore, the total genetic diversity in a population may diminish over time owing to genetic drift.

- ❑ **Founder Effect:** When a novel population is founded, including only a small group of individuals (founders), the genetic makeup of this new population may not represent the genetic diversity of the original population. This is called the "founder effect."

- ❑ **Population Bottleneck:** Sometimes, populations go through a "bottleneck" due to unfavorable conditions or events that remarkably slash the population's size. Even provided the population later recovers in numbers, genetic drift may have already significantly altered the allele frequencies, resulting in reduced genetic diversity.

- ❑ **Homogeneity *vs.* Diversity:** In summary, genetic drift tends to make small populations more homogeneous by causing the elimination or fixation (making them the only variant) of certain genes. In contrast, larger populations tend to maintain greater genetic diversity.

In practical terms, the relevance of genetic drift in implications for the genetic health and adaptability of populations, notably in the context of conservation efforts and understanding how genetic diversity is impacted by factors like habitat destruction or small founder populations.

6.6. Concept of Population Improvement

In plant breeding, when we intend to enhance open-pollinated or cross-pollinated species, the main striving to change the genetic makeup of the population by favoring specific desirable traits while keeping a good amount of genetic diversity. Unlike self-pollinated species, where we focus on individual plants and aim for uniformity, population improvement looks at the entire group of plants, and they aren't all the same.

6.6.1. Types

Two key methods can be employed to improving open-pollinated populations in plant breeding: population improvement and the development of synthetic cultivars.

- ❐ **Population Improvement:** This method involves changing the entire population by using specific selection strategies. Cultivars developed this way can sustain their identity through random mating within the group over time.

- ❐ **Synthetic Cultivars:** These are open-pollinated cultivars created by combining inbred or clonal parental lines. Despite this, these cultivars are not sustainable and must be recreated from the parental stock.

6.7. Methods of Population Improvement

Before selecting plants, an evaluation process takes place. The selection of breeding material follows an assessment of the available variability. For self-pollinated species, individuals are homozygous, and their genetic makeup is precisely replicated in their offspring, making a progeny test suitable for evaluating an individual's performance.

However, open-pollinated species are made up of heterozygous plants, and they are further cross-pollinated by other heterozygous plants in their vicinity. Progeny testing is not sufficient for assessing individual plant performance in such species. To get a more accurate evaluation, plant breeders use pollen, preferably from a homozygous source (like an inbred line), to pollinate the plants.

One way to evaluate the performance of different mother plants comparatively, using a common pollen source (tester line), is called a "test cross." This test aims to assess how well a parent performs in a cross, which is known as combining ability.

Plant breeders use two main categories of methods in population improvement, based on how they evaluate performance:

- ❐ **Phenotypic Selection:** This method relies on observing and selecting plants based on their physical characteristics or traits.

- ❐ **Progeny Testing (Genotypic Selection):** Progeny testing involves evaluating plants reflecting the performance of their offspring. It's especially useful for self-pollinated species.

Specific methods within these categories include mass selection, half-sib, full-sib, recurrent selection, and synthetics. These methods help breeders improve plant populations by selecting the best individuals based on their traits and genetics.

6.8. Practical Implementation

The practical implementation of genetic knowledge and population structure in conserving plant diversity involves collecting and preserving plant germplasm (seeds or reproductive material). It is described as follows

6.8.1. Collecting Germplasm

Gene banks collect plant germplasm when genetic diversity is at risk, or when there's a need to fill gaps in diversity. Collection sites include farmers' fields (especially for traditional varieties), gardens, markets, and wild habitats. Knowledge of species distribution, breeding systems, and population structure helps in planning collection.

6.8.2. Genetic Drift Consideration

Genetic drift is the random change in allele frequencies in small populations. It's crucial to consider when collecting germplasm. A sample of about 172 plants is usually enough to preserve most genetic variation. Each plant's seeds should be stored separately, not mixed.

6.8.3. Regeneration Strategy

When regenerating seeds from germplasm, it's important to maintain genetic diversity. For outcrossing species (those that cross-pollinate), a paired crossing design is used. This solidifies that each individual contributes two alleles to the next generation, reducing genetic drift.

6.8.4. Conserving Crop Wild Relatives

For in situ conservation (conservation in their natural habitat), the minimum viable population size is typically around 5000 individuals. Ideally, consideration should be given to having several populations per species conserved. This "50/500 rule" helps maintain genetic diversity.

6.8.5. On-Farm Conservation

On-farm conservation focuses on preserving traditional crop varieties (landraces) in their natural environment. Landraces often reveal heightened genetic diversity as a result of local adaptations and farmer selection.

In summary, genetic knowledge of population guides the collection and preservation of plant germplasm to ensure the conservation of genetic diversity. Careful strategies, like maintaining separate seeds and using specific regeneration techniques, help prevent genetic drift and maintain diversity in gene banks. For in situ conservation, multiple populations of a species are conserved to ensure genetic robustness. On-farm conservation keeps traditional crop varieties in their natural context to maintain genetic diversity.

6.9. Conclusion

In conclusion, plant population genetics is imperative for fathoming how plants modify to their conditions and evolve in answer to changing environments. It not only informs scientific inquiry but also encompasses practical applications in plant breeding, conservation, and ecosystem management. The foundational principles of population genetics guide us in preserving endangered species and maintaining genetic diversity. As genetic research advances, this knowledge continues to empower us to make informed decisions in breeding programs, ensuring the health and diversity of plant populations. In essence, plant population genetics is a vital field that bridges science and conservation, guiding us toward a better understanding and protection of our botanical world's genetic diversity.Top of Form

References

Falconer, D.S. (1960) Introduction to Quantitative Genetics. Oliver and Boyd Ltd., Edinburgh, 1-140.Top of Form

Lawrence, M.J. and Marshall, D.F. (2000). Plant population genetics. In: Maxted, N., Ford-Lloyd, B.V., Hawkes, J.G. (eds) Plant Genetic Conservation. Springer, Dordrecht.

Doolittle, D.P. (2012). Population Genetics: Basic Principles (Vol. 16). Springer Science & Business Media, London.

Chapter–7

Pan Genomics: Leveraging Wild Relatives for Crop Improvement

Himani P. Vadodariya[1], Hensi D. Kundaria[2] and Chirag P. Chandramaniya[3], Purnima Ray[4*]

2,3 M.Sc. Student, Department of Genetics and Plant Breeding, NAU, Navsari, Gujarat

1,4 Ph.D Student, Department of Genetics and Plant Breeding, NAU, Navsari, Gujarat

**Corresponding author: purnimaroy36@gmail.com*

7.1. Introduction

The plant genome defined as a complete set of genetic material present within the cells of a plant species. Genomes contain all the information necessary for an organism's growth, development, functioning, and reproduction. Plant genomes, which is having high complexity along with diversity was extensively studied for years[1]. Research on plant genome is critical for increasing crop yields, producing disease-resistant cultivars, studying plant response to changing environments and even progressing biotechnological applications.

Advances in DNA sequencing technology have helped scientists to interpret the full DNA sequence of various plant species, considerably improving our understanding of plant biology, evolution, and variety. Because of these new advancements, around 1000 accessible genomes of plants belongs to 800 species are widely distributed among different plant families[2]. Beyond these technological advancements and initiatives, the scientists are also dedicated to sequencing individuals of same species that reveal high intraspecific diversity in plants, these finding can even surpass the divergence found between humans and chimpanzees.[3].

Dynamic plant genomes exist in and within various populations of plants as a result of several rounds of genomic variation, duplication, intragenic and intra specific recombination and subsequent fractionation. In addition to that, tandem duplication, transposable elements, deletion and plant rearrangement also play a great role in plant genome variation across populations. These processes can result in gene presence/absence variation (PAV) and structural variation (SV), both of which are linked to phenotypic variety and adaptation as well as selection. As a result of this variation in genes, single reference genome assemblies do not accurately represent the genomic diversity within a species, which demand a more expansive view of genetic variation. Consequently, a pangenome is necessary to capture this variety. The pan-genome is thought to be more accurate and comprehensive than the often used single reference genome of many species[4], since it represents the entirety of DNA sequence variability within a species[5].

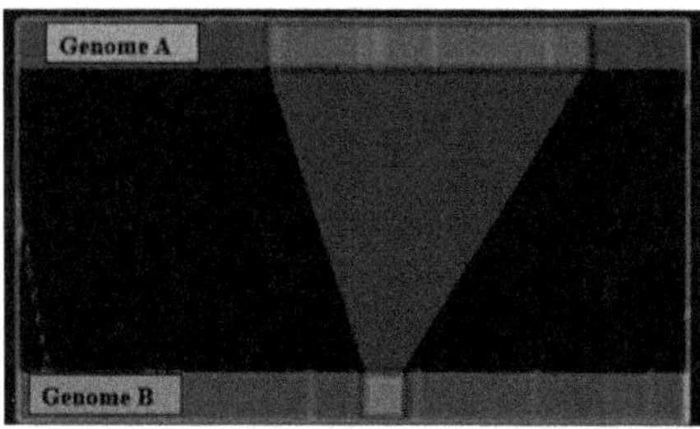

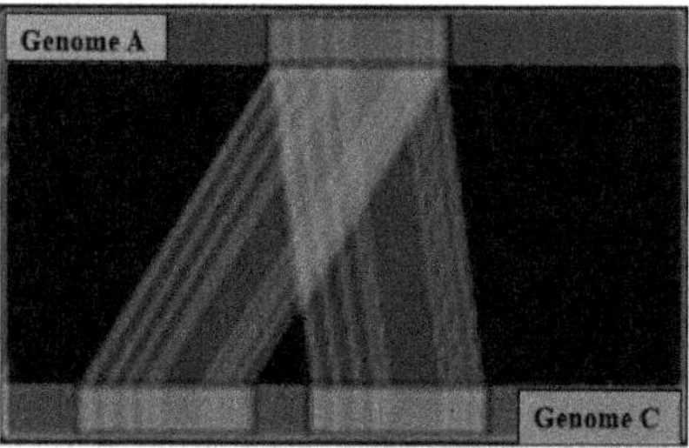

Presence/absence variantCopy number variant

Figure 7.1: Representation of presence/absence variants (PAVs) and a copy number variants (CNVs) structural variants for Genome A

7.2. Need of a Pan Genome

Inadequate understanding of the range, occurrence and processes behind a practically alignment of sequence reads from individual plant species to actual reference genome successfully detects micro polymorphisms. These includes Single Nucleotide Polymorphisms (SNPs) and smaller insertions and deletions (indels), but completely overview more than 50 bp longer segments of newer sequences that are found absent from the reference genome and sequences that greatly deviate from the actual reference genome. Plant pan-genomes were created as a result of plant variations, some of that larger structural variants include both the Presence or Absence Variants (PAVs) and a Copy Number Variant (CNVs) (Figure 1). Sometimes, structural variants contain many genes, while numerous studies also show the crucial impact of structural variants on qualities that are crucial to agriculture (e.g., resistance to pest, diseases, drought, salinity etc, plant morphology, days to flowering, plant physiology, economic yield and qualities). The creation of plant pan-genomes was prompted by a lack of understanding regarding the spectrum, prevalence, and reasons for the emergence of structural variations in plants.

Basically, the pan genome is referred as a non redundant set of genes. This definition is based on gene, which misses a diversity of the noncoding intron sequences and some repetitive sequences like a transposable element. Many of these structural variants, including inversions and translocations, might be significant for gene regulations and phenotypic traits in plant species. A more precise definition of pan genome is intraspecific non redundant set of the genetic sequences (Figure 2C), which is challenging to construct, display and interpret as compared to pan genome based on a gene. Due to several technical constraints, the plant pan genomes that have been created to far are mostly based on genes. Recent advances in DNA sequencing, on the other hand, are significantly lowering the expenses of genomic assembling and reducing some of the hurdles to create effective plant pan genome.

The prefix "pan" from pan genome derives from a Greek word pan, which means "all". Although, pan genome analysis did not reach in a higher eukaryote until 2010, when high throughtput next generation sequencing of short reads was developed and *de novo* assembly in higher organisms were successful. Comparison of *de novo* assemblies and annotations from eight different strains of the bacterial species (*Streptococcus agalactiae*) leads to development of the first pan-genome in 2005[6]. Specifically in a plant, first attempt at pangenome assembly was done in soybean[7], which further extended the pan-genome research to the other plant species. Then onwards a closed pan genome was observed in maize[8], *B. oleracea*[9], wheat[10], rice[11] and *B. napus*[12]. This success has given new wing to a study of crop diversity and improvement. Study of pan genome on these crops have reveled unexpected higher genomic variability and diversity.

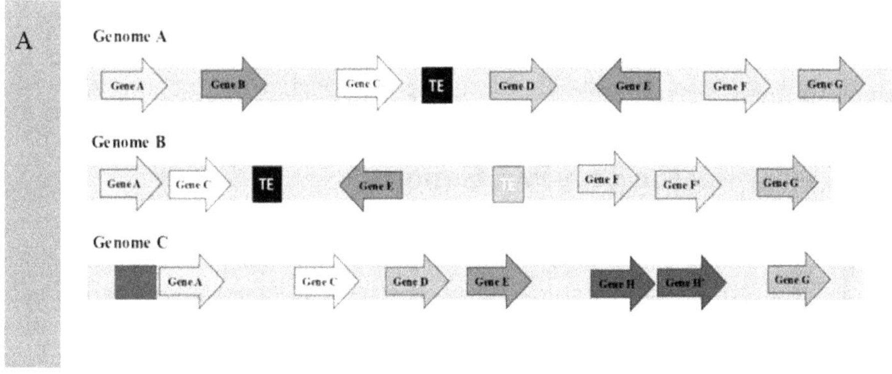

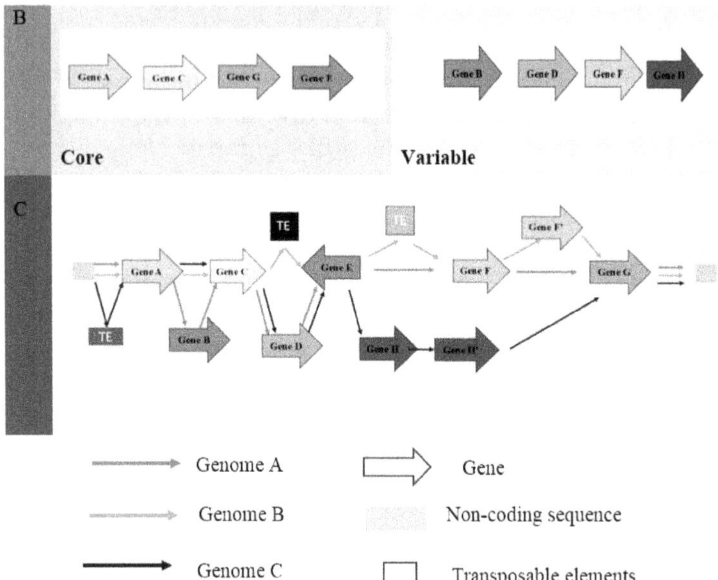

Figure 2: Diagramatic representaton for construction of a pan genome (A) Syntenic genomic segments from different three genomes i.e. Genome A, B and C. For the sake of clarity, introns and minor polymorphisms (SNPs and indel) have been left out. Note that Gene F and H includes tandem duplicates, Gene E is inverted in genome 1 and 2. (B) Core genes and variable genes among three genomes (C) Graphical representation of sequence based pan genome of genome A, B and C. The genomic sequences of any individual genome can be reconstructed from Figure C by following the path of colored arrows: Colour Red represents genome A, Green represents genome B, and Blue repre sents genome C.

Thus, it can be concluded that the plant genome is a blueprint of genetic information which tends to inherent in a specific plant species and serves as the foundation for understanding its biology and history of evolution. However, the concept of a pan-genome broadens this perspective by encompassing the collective genomic diversity present within a species.

7.3. Classification of a Pan Genome

A pan genome can be defined as a complete set of individual genes and genetic elements found in a group of related organisms, which may include distinct strains or species. On the basis of whether Pan-genomes are evolving or not, it is classified into two categories, each with its unique set of characteristics:

7.3.1. Open pan genome

As more strains or individuals are being sequenced, the open pan-genome will continue to grow. That means, each new sequenced genome adds additional genes and genetic elements to the pan-genome as a whole. Generally, Bacterial species have open pan-genomes, which contribute to continuing genome growth through a horizontal gene transfer and genetic variability by genetic expansion. E. coli genome size is in the 4000-5000 gene range, and the pangenome size for this species is predicted to be around 2000 genomes, with 89,000 distinct gene families[13].

7.3.2. Closed pan genome

A closed pan genome is one with a very well-defined gene diversity that is unlikely to change dramatically with the addition of new genomes. Humans and other eukaryotic species have a closed pan genome.

Generally, construct of a pan genome is divided in to two parts: core and dispensable (Figure 2B).

7.3.2.1. Core genome

The core genome from a complete pan genome consists of those genes, which are present in all individual strains or species. The ratio of non-synonymous mutation to the synonymous mutations is lower in the core genome set as compare to the variable or dispensable gene set of *B. distachyon*[14], *B. napus*[12] and soybean[7]. Core genome shown a higher expression level across the tissues than variable genome and connected with important cellular functions. Generally, core genome was found to be larger on average than variable genes.

7.3.2.2. Variable and dispensable genome:

The dispensable or variable genome includes those genes, which are absent from one or more genomes. This genome is tend to be more variable than the core genome. Additionally, higher single nucleotide polymorphism and insertin-deletion density has been also observed in the variable genome set as compared to the core genome. There is a higher proportion of variable genes (without functional annotation) were found in soybean (58% versus 34%)[7] and rice (60% versus 37%)[15] as compared to the core genome. Moreover, the higher percentage of variable genome without presence of any homologs in other plant species were also found in soybean i.e. 17% versus 5%[7]. This genome is being connected with a function related to environmental and defense related responses, receptor related activities, signal transduction, antioxidant activity and regulation of a gene. Pan genome approach is successfully used to identify genomic regions associated with seed oil content in sesame[16].

Further, the variable genome is divided into two parts: first is an accessory gene, which are found in at least two individual or absent from the at least one strain and the second one is a unique gene, which is found in only one genome or strain.

Thus, core genome tends to be conserved among the species, while variable genome contributes to the species' diversity by allowing it to adapt as per the varied environmental conditions. The dispensable genome contributes significantly to the observable variation of agronomically relevant characters and have a special importance for boosting crop yield. It is also have an important role in adaptive evolution of spices and in a domestication event.

7.4. Construction and Interpretation of Pan Genome

Creating a pan genome involves the compilation and analysis of genetic information from multiple individuals within a species to capture the full genetic diversity of that species. The concept of a pan-genome acknowledges that there is significant sequence variation beyond   the reference genome of a single individual. This can be particularly relevant for species with high levels of genetic diversity, such as plants and certain microbial communities. Unlike small nucleotide polymorphisms and other InDels, which are easily detected and catalogued, effectiveness of pan genome associated with complex genetic variants is remains difficult, even with a use of computer and better genomic technology.[4].

General steps to create pan genome include data collection and sequencing, genome assembly, variant calling and genotyping, construction of pan genome, annotation and analysis followed by visualization.

The initial strategy to construct pan genome relied heavily on parallel read mapping and an assembly to investigate the complexity associated with a plant pan genomes, later ultrahigh-quality precise genomic sequences were used to study variability as completely as possible. However, the genome graph technique is now in use, which makes it simpler to employ pan-genomes by using unified coordinate systems.[17]. In 2020, soybean was used to create the first graph based plant pan genome.[7], It is now anticipated to be a superior alternative to the traditional and linear plant pan genome approaches for more effective analysis.

Thus, it can be briefly said that pan genome can be constructed by

❏ Short read mapping

❏ High quality genome assembly and comparison (Map to Pan)

❏ Pan genome graph

Earlier, attempts were made to use short sequencing reads to detect genome-wide structural changes (insertion, deletion, duplication, and inversion). Genotype of plants are different among spices and also among individual of same species, but genome-wide identification of the structural variations was remained a challenge until the development of a high throughput short-read sequencing technologies[18]. However, it was unable to detect reasonably larger han 30 bp insertions and other complex structural variations. To find that, a series of computer algorithms was created for short read alignment to a reference genome. To identify that read depth approach[19] and read pair methods [20] were used.

Other advanced tool Pindel [21] used a split read strategy to align a split reads to discontinuous genomic areas traversed by putative structural variants. methods have been developed to transformed reads into short k-mers and to identify k-mers that are associated with particular variations.

Some approaches first condense these enormous data into relatively long contigs of several bases to reveal the majority of structural variations in plants and improve the ability of short reads to detect variants, which improve the sensitivity and accuracy with a reduced false discovery for variant calling, additionally, it also helps to detect minor structural variations.

A more streamlined version of the "map-to-pan" technique was created to boost pan genome productivity by mapping all short reads with a the reference genome and then concurrently assembling remaining unmapped reads from different individuals. A "Map-to-pan" is likely one of the most prevalent approaches for successfully building pan genomes in a variety of crops like rice, wheat, tomato, maize, sorghum and cotton. This short read assemblies generally focus on a genic region, while non genic region remained untouched or remained incomplete, which is a major limitation of short read mapping. To counter that, third generation sequencing focus on long reads. PacBio and Nanopore technologies gives a new wing to a genome assembly. These technologies generated a contigs with N50s up to several bases that gives the high quality non genic regions along with interspaced exons.

To the date, several genome assembly tools have been developed, which includes MECAT[22], HiCanu[23], NECAT[24], Hifiasm[24], wtdbg2[25], NextDenovo[26] and Shasta[27]. A whole genome alignment (WGA) across multiple high-quality genomes enhanced the accuracy and sensitivity of structural varient calling. For ease, certain aligner like, Mummer [28] and AnchorWave[29] also have been developed to arrange plant genomes of large sizes with more repetitive sequence. However, because to the occurrence of paralogous sequences in plants, which may readily induce mis-alignment, effective identification of more complicated structural variants like inversion, duplications, and translocations is problematic. The pan genomes created using the "map-to-pan" approach have been attached to the backbone genomic sequences for complete genetic analysis and as a result, such pan genomes have faced the limitation of representation of a single haplotype for each locus with structural variation. So, as alternative, structural variants and their alternative sequences are depicted as a node and edges in a graphical approach. Thus, providing a basic system for the cross comparisons among them.

To create a pan genome with a graph approach, there are two major steps: the first step is storing and indexing the graph and then alignment of reads against that [4]. Genome graphs provide distinct aids in storing complex fixed structural varients such as insertions and deletions with single nucleotide polymorphisms and multiple insertions from the single locus. A better option is to use a unique and different colours along the critical nodes to represent paths with in a population, ensuring that every path with in a graph are corresponding to specific colour. A de Bruijn graph-based method that has previously been employed in assembly and genotyping. However, the computation expenses of analysing a "coloured" genome graph, on the other hand, remain quite high.

To the date, a tool Vg has been used to generate all graph genomes reported in plants[30], which is a one of the most advanced genome graph programmes capable of incorporating complex structural variants, simple small nucleotide polymorphisms and small InDels.

The main limited phases in the analysis of graphs that are based pan-genomes are building and indexing. Once the graph is established, reads may typically alinged to graph reference more quickly than a traditional linear genome. HISAT2 is a main aligner, which has been used for the alignment of reads with a genomic graph by use of different algorithms [31].

7.5. Application of Pan Genome in Crop Improvement

Over the past two decades, there has been a significant advancement in crop breeding, partly due to the full integration of genomic resources and technology into Marker Assisted Selection (MAS), gene pyramiding, Genomic Selection (GS) and genome editing procedures. To generate enough food for the fast expanding global population, crop breeding strategies must be continuously evolving. Pan-genomes provide a number of benefits over conventional single reference genomes[5] and used more frequently in crop genetic research and molecular breeding.

7.5.1. Discovery of Novel Genes and Alleles

Pan-genomic studies help to identify novel genes and alleles that are absent or underrepresented in a reference genome. This can lead to the discovery of important and valuable traits for crop improvement. For example, in a study in maize, highlighted the importance of a pan-genome approach for understanding gene presence/absence variations[32].

7.5.2. Association Breeding

Association mapping utilizes historical recombination in natural to detect a associations between genotypic markers and phenotypic expression. Genotyping a population by complete alignment of short reads with a reference genome by considering historical recombination is a most critical step in association mapping. When a pan-genome is used as the reference genome, reads coming from those sequences that are missing from the reference genome are less likely to misalign and more chances to detect genetic variation in the variable sequences. The pan-genome also offer opportunity for enhancing genotyping in other populations, particularly those that have many genetically distant parents, including MAGIC and NAM populations.

7.5.3. Trait Mapping and QTL Identification

Pan-genome studies enable more accurate trait mapping and identification of quantitative trait loci (QTLs) by capturing allelic variations across diverse accessions. Pan genome approach was used to identify genomic regions for seed oil content in sesame [16].

7.5.4. Epigenetic Studies

It has been extensively studied in plants how epigenetic properties like as DNA methylation, histone modification and small noncoding RNAs demonstrate substantial specificity to a cells and tissues, which is associated with domestication and breeding strategies. However, these studies may be impacted by mapping bias caused by mapping short reads from a genetically distant individual with a reference genome. These ultimately resulted in biased comparison. Use of "pseudo" reference in alternative to a reference genome can mitigate this problem without shifting the genome coordinates.

7.5.5. Functional Annotation and Gene Prediction

Pan genome analyses contribute to better functional annotation of genes by integrating information from multiple accessions.

7.5.6. Marker Development for Breeding

Pan-genomic studies facilitate the development of molecular markers linked to important traits, aiding marker-assisted selection (MAS) in crop breeding.

7.5.7. Resilience and Adaptation

Understanding how various gene sets contribute to a species' resistance and adaptation to various environmental situations, such as drought, pests, and diseases, can be learned via the study of the pan-genome.

Additionally, pan genome is employed in genomic editing, gene pyramiding, marker assisted breeding (MAS), molecular based breeding, genomic selection (GS) and other technologies to enhance important agronomic qualities by serving as a complete reference.

7.6. Challenges and Future Directions

One of the major challenges in pan-genomics is the complex computational and analytical demands required to process and interpret massive amounts of genomic data from multiple individuals or populations. For researchers, integrating diverse datasets that may include various genomic variations, structural variations, and epigenetic modifications is a difficult task. To accurately analyse and interpret this complex genomic data, it is essential to create reliable algorithms and scalable computational frameworks. Additionally, to ensure ethical and open research practices in the pan-genomics era, secure data-sharing platforms and strict data protection regulations must be developed. These concerns about ethics and privacy arise when managing and sharing large-scale genomic datasets from various sources.

Looking ahead, pan-genomics has the potential to make transformative advances. Because of a significant portion of the gene diversity in a wild species has been lost and overall higher rate of variation in a crop species. As a result of domestication, new strategies are needed to identify and fully exploit previously untapped alleles from wild relatives in order to sustain genetic gain. Understanding how genetic variations affect cellular functions and phenotypic outcomes will be made possible by integrating pan-genomic data with other omics datasets like transcriptomics, proteomics, and metabolomics. Single-cell pan-genomics will also reveal complex cellular heterogeneity and dynamic changes within populations as single-cell technologies advance. By accepting these obstacles and looking towards the future, the pan-genomics area is set to revolutionise our understanding of genomics and its applications in several scientific fields.

7.7. Conclusion

In conclusion, pan-genomics has emerged as a revolutionary approach in the field of crop improvement, offering profound insights and transformative opportunities. Through the comprehensive analysis of genetic diversity within a species, pan-genomics has provided a holistic understanding of genomic variations, structural differences, and functional elements that underlie the vast genetic reservoir of crops.

References

1. Danilevicz, M. F., Fernandez, C. G. T., Marsh, J. I., Bayer, P. E. and Edwards, D. Plant pangenomics: approaches, applications and advancements. Current opinion in plant biology. 2020; 54: 18-25.

2. Sun, Y., Shang, L., Zhu, Q. H., Fan, L., & Guo, L. Twenty years of plant genome sequencing: achievements and challenges. Trends in Plant Science. 2022.

3. Buckler, E. S., Gaut, B. S., & McMullen, M. D. Molecular and functional diversity of maize. Current opinion in plant biology. 2006; 9(2): 172-176.

4. Bayer, P. E., Golicz, A. A., Scheben, A., Batley, J., & Edwards, D. Plant pan-genomes are the new reference. Nature plants. 2020; 6(8): 914-920.

5. Sherman, R. M., & Salzberg, S. L. Pan-genomics in the human genome era. Nature Reviews Genetics. 2020; 21(4): 243-254.

6. Tettelin, H., Masignani, V., Cieslewicz, M. J., Donati, C., Medini, D., Ward, N. L. et al. Genome analysis of multiple pathogenic isolates of Streptococcus agalactiae: implications for the microbial "pan-genome". Proceedings of the National Academy of Sciences. 2005; 102(39): 13950-13955.

7. Li, Y. H., Zhou, G., Ma, J., Jiang, W., Jin, L. G., Zhang et al. De novo assembly of soybean wild relatives for pan-genome analysis of diversity and agronomic traits. Nature biotechnology. 2014; 32(10): 1045-1052.

8. Hirsch, C. N., Foerster, J. M., Johnson, J. M., Sekhon, R. S., Muttoni, G., Vaillancourt, B. et al. Insights into the maize pan-genome and pan-transcriptome. The Plant Cell. 2014; 26(1): 121-135.

9. Golicz, A. A., Bayer, P. E., Barker, G. C., Edger, P. P., Kim, H., Martinez, P. A. et al. The pangenome of an agronomically important crop plant Brassica oleracea. Nature communications. 2016; 7(1): 13390.

10. Montenegro, J. D., Golicz, A. A., Bayer, P. E., Hurgobin, B., Lee, H., Chan, C. K. K. et al. The pangenome of hexaploid bread wheat. The Plant Journal. 2017; 90(5): 1007-1013.

11. Wang, W., Mauleon, R., Hu, Z., Chebotarov, D., Tai, S., Wu, Z. et al. Genomic variation in 3,010 diverse accessions of Asian cultivated rice. Nature. 2018; 557(7703): 43-49.

12. Hurgobin, B., Golicz, A. A., Bayer, P. E., Chan, C. K. K., Tirnaz, S., Dolatabadian, A. et al. Homoeologous exchange is a major cause of gene presence/absence variation in the amphidiploid Brassica napus. Plant biotechnology journal. 2018; 16(7): 1265-1274.

13. Land, M., Hauser, L., Jun, S. R., Nookaew, I., Leuze, M. R., Ahn, T. H. et al. Insights from 20 years of bacterial genome sequencing. Functional & integrative genomics. 2015; 15, 141-161.

14. Gordon, S. P., Contreras-Moreira, B., Woods, D. P., Des Marais, D. L., Burgess, D., Shu, S. et al. (2017); Extensive gene content variation in the *Brachypodium distachyon* pangenome correlates with population structure. Nature communications, *8*(1), 2184.

15. Schatz, M. C., Maron, L. G., Stein, J. C., Wences, A. H., Gurtowski, J., Biggers, E et al. (2014); Whole genome de novo assemblies of three divergent strains of rice, *Oryza sativa*, document novel gene space of aus and indica. Genome biology, *15*, 1-16.

16. Yu, J., Golicz, A. A., Lu, K., Dossa, K., Zhang, Y., Chen, J. and Zhang, X. (2019); Insight into the evolution and functional characteristics of the pan□genome assembly from sesame landraces and modern cultivars. Plant Biotechnology Journal, *17*(5), 881-892.

17. Wang, T., Antonacci-Fulton, L., Howe, K., Lawson, H. A., Lucas, J. K., Phillippy, A. M. Human Pangenome Reference Consortium. The Human Pangenome Project: a global resource to map genomic diversity. Nature. 2022; 604(7906): 437-446.

18. Metzker, M. L. Sequencing technologies the next generation. Nature reviews genetics. 2010; 11(1): 31-46.

19. Zhou, Z., Jiang, Y., Wang, Z., Gou, Z., Lyu, J., Li, W. et al. Resequencing 302 wild and cultivated accessions identifies genes related to domestication and improvement in soybean. Nature biotechnology. 2015; 33(4): 408-414.

20. Ho, S. S., Urban, A. E., & Mills, R. E. Structural variation in the sequencing era. Nature Reviews Genetics. 2020; 21(3): 171-189.

21. Ye, K., Guo, L., Yang, X., Lamijer, E. W., Raine, K.and Ning, Z. (2018); Split-read indel and structural variant calling using PINDEL. Copy Number Variants: Methods and Protocols, 95-105.

22. Xiao, C. L., Chen, Y., Xie, S. Q., Chen, K. N., Wang, Y., Han, Y. MECAT: fast mapping, error correction, and de novo assembly for single-molecule sequencing reads. nature methods. 2017; 14(11): 1072-1074.

23. Nurk, S., Walenz, B. P., Rhie, A., Vollger, M. R., Logsdon, G. A., Grothe, R. et al. HiCanu: accurate assembly of segmental duplications, satellites, and allelic variants from high-fidelity long reads. Genome research. 2020; 30(9): 1291-1305.

24. Chen, Y., Nie, F., Xie, S. Q., Zheng, Y. F., Dai, Q., Bray, T. et al. Efficient assembly of nanopore reads via highly accurate and intact error correction. Nature Communications. 2021; 12(1): 60.

25. Scott, M. F., Ladejobi, O., Amer, S., Bentley, A. R., Biernaskie, J., Boden, S. A. et al. Multi-parent populations in crops: a toolbox integrating genomics and genetic mapping with breeding. Heredity. 2020; 125(6): 396-416.

26. Hu, J., Fan, J., Sun, Z., & Liu, S. Next Polish: a fast and efficient genome polishing tool for long-read assembly. Bioinformatics. 2020; 36(7): 2253-2255.

27. Shafin, K., Pesout, T., Lorig-Roach, R., Haukness, M., Olsen, H. E., Bosworth, C. Nanopore sequencing and the Shasta toolkit enable efficient de novo assembly of eleven human genomes. Nature biotechnology. 2020; 38(9): 1044-1053.

28. Marçais, G., Delcher, A. L., Phillippy, A. M., Coston, R., Salzberg, S. L., & Zimin, A. MUMmer4: A fast and versatile genome alignment system. PLoS computational biology. 2018; 14(1): e1005944.

29. Song, B., Marco-Sola, S., Moreto, M., Johnson, L., Buckler, E. S., & Stitzer, M. C. AnchorWave: Sensitive alignment of genomes with high sequence diversity, extensive structural polymorphism, and whole-genome duplication. Proceedings of the National Academy of Sciences. 2022; 119(1): e2113075119.

30. Garrison, E., Siren, J., Novak, A. M., Hickey, G., Eizenga, J. M., Dawson, E. T. et al. Variation graph toolkit improves read mapping by representing genetic variation in the reference. Nature biotechnology. 2018; 36(9): 875-879.

31. Kim, D., Paggi, J. M., Park, C., Bennett, C., & Salzberg, S. L. Graph-based genome alignment and genotyping with HISAT2 and HISAT-genotype. Nature biotechnology. 2019; 37(8): 907-915.

32. Jiao, Y., Peluso, P., Shi, J., Liang, T., Stitzer, M. C., Wang, B. et al. Improved maize reference genome with single-molecule technologies. Nature. 2017; 546(7659): 524-527.

Chapter–8

Tools and Techniques for Exploiting Genomic Diversity in Crop Improvement

Hari Haran R., Ganesh Babu SRVN., Karunya N[*]

**Corresponding author: karunyastar22@gmail.com*

8.1. Introduction

The relentless expansion of the world's population is on an unyielding trajectory, presenting an imminent and formidable challenge: a substantial surge in global demand for essential crops. Simultaneously, plant breeders worldwide face an ever-mounting challenge. They must continuously enhance crop varieties that not only deliver high and consistent yields but can also thrive in the face of diminishing agricultural space and the increasingly harsh conditions brought by climate change. The unequivocal confirmation of global warming by the Intergovernmental Panel on Climate Change (IPCC) underscores that climate change is a perilous risk factor. It threatens to curtail crop output due to the proliferation of unfavorable climatic conditions. In this critical juncture of agriculture's evolution, modern crop varieties exhibit reduced genetic variation and skewed allele frequency spectra compared to their wild counterparts. This reduction arises from the rigorous human-mediated selection that characterizes plant breeding efforts. Meeting this daunting challenge necessitates a profound examination of the genetic diversity within our crop species. Indeed, genetic diversity emerges as the linchpin of success in breeding programs. It furnishes breeders with the fundamental groundwork for selecting cultivars that exhibit superior agricultural performance. The contemporary era has bestowed upon us a powerful arsenal of genomics tools that were hitherto unimaginable. These tools provide an unprecedented platform for accelerating crop improvement and amplifying genetic gains in breeding programs. They are especially invaluable for traits that are arduous, time-consuming, or exorbitant to assess phenotypically.

In this era of genomics, breeders are equipped with the capacity to swiftly and economically profile the genomes of even vast plant populations. This is made possible through ultra-high-throughput genotyping techniques such as Genotyping-by-Sequencing (GBS) and array-based Single Nucleotide Polymorphism (SNP) genotyping. These techniques yield comprehensive, high-resolution molecular data. Breeders can now glean in-depth and exceedingly precise insights into genetic diversity at the chromosomal and sub-genomic levels. By doing so, they can unearth genes linked to beneficial biological functions. Through the identification and harnessing of genes associated with advantageous traits, and by judiciously incorporating these genetic elements into breeding programs, we can usher in a new era of crop varieties that can withstand the challenges posed by a growing population and a changing climate.

8.2. Importance of Genetic Diversity

Genetic diversity is a cornerstone of agricultural progress, playing a pivotal role in enhancing crop production, ensuring adaptability to changing environmental conditions, and fostering sustainable agricultural practices. This diversity is derived from the vast genetic resources existing across the globe and encompasses a wide range of plant species, including cultivars, landraces, and wild specimens. In the pursuit of increased food production and sustainable agriculture, genetic diversity stands as an indispensable resource. Over time, domestication and evolution have led to the loss of numerous beneficial alleles, rendering them untapped within the genetic makeup of cultivated crops. However, these hidden treasures of genetic variation are still preserved within the wild relatives and landraces of crop plants, holding the potential to drive the development of superior agronomic cultivars.

To harness this potential, the initial step in any research endeavor targeting beneficial genes in crops is the thorough examination of genetic and phenotypic diversity within available genetic resources. Understanding the genetic underpinnings of traits, such as population diversity, population structure, and linkage disequilibrium, is essential. This understanding is facilitated by the analysis of genome-scale and population-scale DNA polymorphisms, unveiling the genetic landscape of the species under investigation.

8.3 Techniques for Exploiting Genomic Diversity

8.3.1 Advances in Next Generation Sequencing (NGS) for Crop Genomics

Next Generation Sequencing (NGS) technologies have revolutionized the study of genetic diversity in crops by enabling rapid screening of millions of single nucleotide polymorphisms (SNPs) across entire genomes. These high-throughput sequencing methods have significantly reduced the cost and time required for genome analysis. Over the past 15 years, NGS platforms have provided reference sequences for complex crop genomes, including large genomes like bread wheat. High-density genic polymorphism data generated by NGS have been applied in quantitative trait dissection, marker-assisted breeding, genomic selection, and exploration of genetic resources (Bolger *et al.,,* 2014). NGS technologies fall into two categories: second and third generation, with the latter offering deeper analysis and more data (Qian *et al.,,* 2014; Voss-Fels *et al.,,* 2015). Second-generation platforms, such as 454 Life Sciences Roche, SOLiD, and Illumina, played a pivotal role in advancing sequencing techniques. Ion Torrent, based on sequencing by synthesis, further reduced costs. Illumina is widely used for genotyping studies in crops due to its versatility and outputs (Ganal *et al.,,* 2012). Third-generation platforms, like Pacific Biosciences (PacBio) and Oxford Nanopore Technologies (ONT), offer long-read sequencing, increasing accuracy and enabling better assembly in repetitive regions. MinION, a nanopore sequencer, produces long reads and is highly portable (PacBio; ONT) (Huang and Han, 2014).

Whole-genome re-sequencing, either with or without a reference sequence, has provided unbiased genetic diversity information. Reduced-representation genotyping-by-sequencing (GBS) methods have made high-throughput genomic resequencing cost-effective for large plant populations, facilitating detailed genetic diversity and population structure analyses (Varshney *et al.,,* 2014). Exome capture allows cost-effective identification of genetic variations in protein-coding regions (Ganal *et al.,,* 2012). RNA-seq, either with a reference genome or as a de novo method, has become essential for studying gene expression, alternative splicing, and genetic variations (Bolger *et al.,,* 2014). High-density SNP genotyping arrays, generated using NGS data, are valuable for genetic analysis in crops, with platforms like Illumina's Infinium and Affymetrix's Axiom being widely used (Thomson, 2014).

8.3.2 Allele Mining

Allele mining has emerged as a promising approach for tapping into the genetic diversity preserved in gene banks worldwide. It involves the dissection of naturally occurring allelic variation in genes controlling crucial agronomic traits. Allele mining provides insights into the evolution of alleles, the identification of new haplotypes, and the development of allele-specific markers for marker-assisted selection and genetic engineering. While initial studies of allele mining have primarily concentrated on coding sequences or exons of genes, recent reports emphasize the significance of nucleotide changes in regulatory regions and non-coding regions. These non-coding regions, including promoters, introns, and untranslated regions (UTRs), can profoundly influence transcript synthesis and accumulation, thereby impacting trait expression. This broader approach, referred to as promoter mining, sheds light on gene regulation mechanisms and offers the isolation of novel and efficient promoters for genetic engineering applications.

TILLING (Targeting Induced Local Lesions IN Genomes) is another technique that aids in identifying polymorphisms, particularly point mutations, in target genes through heteroduplex analysis. A variant of this approach, Eco TILLING, assesses natural variation in selected genes within crop species. These techniques are potent tools for haplotyping and SNP discovery, contributing to genetic diversity exploration in crops. Incorporating approaches like in vitro and in silico promoter mining alongside traditional allele mining programs is crucial. This integration allows for the identification and categorization of cis-regulatory elements and a deeper understanding of their role in gene regulation. While the potential benefits of allele mining in plant genetic resources management are evident, it's essential to note that many international crop research institutes have initiated efforts to characterize allelic diversity in crop plants. These initiatives align with the overarching goal of unlocking genetic diversity in crops to benefit resource-poor communities and ensure global food security.

8.3.3 Genome-Wide Association Studies (GWAS): Uncovering Genetic Links to Traits

GWAS is a powerful method for investigating the associations between a comprehensive set of single-nucleotide polymorphisms (SNPs) and various phenotypic traits. It involves the examination of genetic variants across the genomes of numerous individuals within a population to identify connections between genotypes and phenotypes. The evaluation relies on the concept of linkage disequilibrium (LD), which involves the non-random association of alleles at different loci. This method has been significantly advanced by the development of high-throughput genotyping technologies. One of the primary purposes of GWAS is to identify genes or loci contributing to specific traits using high-resolution genome-wide markers. It goes beyond merely pinpointing genetic elements linked to traits and offers insights into the complex genetic architecture underlying these traits. In recent years, GWAS has been instrumental in detecting numerous Quantitative Trait Loci (QTLs) and candidate genes associated with various traits in plants like Arabidopsis, rice, wheat, soybean, and maize.

8.3.3.1 Association Mapping and LD

GWAS is a form of association mapping that relies on the concept of linkage disequilibrium. It assumes that a marker locus is sufficiently close to a trait locus for a marker allele to co-occur with the trait allele over multiple generations. This approach helps identify inter-individual genetic variants, primarily SNPs, that are statistically correlated or in LD with causal variants.

8.3.3.2 Comprehensive Genome Analysis

GWAS examines a vast array of common genetic variants across different individuals to identify if any of these variants are associated with a specific trait. This approach is particularly valuable for understanding the genetic basis of major diseases and complex traits.

8.3.3.3 Case-Control Studies

In GWAS, individuals with a particular disease or trait (cases) are compared to similar individuals without the condition (controls). DNA samples are collected, and millions of genetic variants are analyzed using SNP arrays. SNPs more frequent in individuals with the disease are considered "associated" with the condition.

8.3.3.4. GWAS Design

Effective GWAS begins with the collection and recording of the phenotype of interest, which can be quantitative or dichotomous (case-control). Quantitative traits can provide higher statistical power to detect genetic effects, while case-control studies can be effective in identifying multiple genes associated with the phenotype.

8.3.3.5. SNP Capture

SNPs are captured using genotyping technologies like microarray technology or next-generation sequencing. These SNPs are then associated with phenotypes based on the GWAS design.

8.3.3.6. Statistical Analysis

The analysis of GWAS data depends on the study design. For quantitative traits, logistic regression may be used, while case-control designs often involve generalized linear models (GLM) and analysis of variance (ANOVA).

8.3.3.7. Replication and Validation:

It's crucial to replicate GWAS findings across different studies and populations. Consistency in results strengthens the association between SNPs and phenotypes.

8.3.3.8. Manhattan Plot

GWAS results are typically visualized using Manhattan plots, which display P-values on a logarithmic scale. SNPs are plotted linearly based on their chromosomal positions, highlighting significant associations.

8.3.3.9. Contrast with QTL Analysis

While QTL (Quantitative Trait Locus) analysis identifies genomic regions affecting genotypes broadly, GWAS specifically pinpoints individual SNPs linked to traits. GWAS offers high specificity in identifying causal variants.

8.3.4. Omics Technologies in Crop Research: Unlocking Genetic and Metabolic Insights

Over the years, omics technologies have emerged as powerful tools in crop research, enabling a comprehensive understanding of genetic and metabolic processes. These technologies include Epigenomics, Transcriptomics, Proteomics, and Metabolomics, each offering unique insights into the intricate world of crop biology.

8.3.5. Epigenomics: Unlocking Epigenetic Modifications

Epigenomics has evolved into a cutting-edge technology for deciphering changes in gene regulation caused by epigenetic modifications to DNA sequences. These modifications, such as DNA methylation, histone modifications, and chromatin accessibility, provide crucial insights into how crops respond to environmental changes without altering their DNA sequences. Epigenetic research has revealed that drought-tolerant plants often exhibit stable methylomes, with differentially methylated regions (DMRs) connecting to stress response and programmed cell death genes. Similarly, histone modifications like H3K27me3 and H3K27ac dynamically regulate rice gene expression in response to cold stress. Stress signals can induce both gene methylation changes and broader histone methylome alterations, linking genome-wide transcription remodeling to stress-induced histone changes.

8.3.6. Transcriptomics: Profiling Gene Expression

Transcriptomics explores RNA transcripts produced by an organism's genome, encompassing both mRNAs and non-coding RNAs (ncRNAs). Traditional profiling techniques like DD-PCR, cDNAs-AFLP, and SSH have given way to high-throughput sequencing methods like RNA-seq, allowing for a more precise characterization of the transcriptome. Transcriptomics studies have unveiled crucial insights into ion homeostasis in salt-tolerant plants and identified transcription factors like MYB, bZIP, and WRKY in the salt-responsive signaling pathway of maize roots. Moreover, non-coding RNAs, including miRNAs, circRNAs, lncRNAs, rRNAs, and siRNAs, have emerged as prospective targets for crop improvement, offering a wealth of regulatory insights.

8.3.7.Proteomics: Investigating the Proteome

Proteomics delves into the study of all proteins expressed within an organism, categorized into sequence, structure, function, and expression proteomics. This field employs various techniques, such as structural proteomics using NMR and sequence proteomics analyzed via HPLC. Gel-based techniques like SDS-PAGE, 2-DE, and 2D-DIGE, along with protein chips and microarrays, enable the efficient analysis of protein expression. NMR spectroscopy allows for the determination of protein structures, aiding in understanding their biological functions. However, single-cell proteome quantification is still evolving, with mass spectrometry playing a pivotal role in assessing protein levels in single cells.

8.3.8. Metabolomics: Uncovering Metabolic Insights

Metabolomics is a relatively recent omics technology for exploring metabolites and understanding crop resilience. Untargeted metabolome detection accelerates integrated metabolomics in plant research, with secondary metabolites playing a crucial role in complex metabolic cascades. Analytical techniques like liquid chromatography, gas chromatography-mass spectrometry (GS-MS), capillary electrophoresis-mass spectrometry (CE-MS), HPLC, NMR, and direct flow injection (DFI) facilitate metabolite profiling. Combining metabolomics with NGS enables the inference of early metabolic networks from genetic sequences, offering opportunities to develop climate-resilient crops. Integrated analyses have highlighted the importance of amino acid metabolism, ABA signaling pathways, and drought resistance mechanisms in crops.

8.3.9. Diversity Array Technology (DArT): Revolutionizing Genetic Diversity Analysis

Diversity Array Technology (DArT) is a high-throughput genotyping technique widely utilized for assessing genetic diversity in various organisms, including plants, animals, and microorganisms. Developed to provide an efficient and cost-effective method for genotyping a large number of DNA samples, DArT has found applications in genetic diversity analysis, population studies, and marker-assisted breeding programs.

Similar to AFLP procedures, DArT employs microarray-based nucleic acid hybridization instead of gel electrophoresis to detect polymorphism. The DArT analysis involves several key steps:

8.3.9.1. DNA Extraction

The process begins with the extraction of DNA from the target samples, which can include individuals or populations within a species.

8.3.9.2. Restriction Enzyme Digestion

One or more restriction enzymes are used to cleave the DNA at specific recognition sites. The choice of enzymes can be tailored based on research objectives.

8.3.9.3. Ligation of Adapters

Specific adapters are ligated to the DNA fragments following digestion. These adapters contain sequences complementary to the ends of the digested DNA fragments.

8.3.9.4. Polymerase Chain Reaction (PCR) Amplification

DNA fragments with ligated adapters are specifically amplified using PCR, ensuring there is enough DNA for subsequent analysis.

8.3.9.5. Microarray Hybridization

DArT employs microarray technology to assess genetic diversity. A microarray consists of thousands to millions of DNA probes or sequences immobilized on a solid surface. Before hybridization, fluorescent markers are applied to the PCR-amplified DNA from the samples. The degree of hybridization between the labeled DNA and microarray probes indicates the presence or absence of specific DNA markers in the samples.

8.3.9.6. Data Analysis

After hybridization, the microarray is scanned to detect fluorescence signals, indicating the presence or absence of particular DNA markers. The results are then used to determine genetic diversity. Various statistical techniques, such as genetic similarity or dissimilarity calculations, can be applied to assess genetic diversity indices.

8.4. Conclusion

The elucidation of genetic diversity relies significantly on a spectrum of sophisticated scientific tools and techniques. Cutting-edge methodologies, such as next-generation DNA sequencing, gene editing and bioinformatics, are instrumental in characterizing genetic variations at unprecedented scales. These tools facilitate the identification of genetic markers, assessment of population structures, and exploration of genomic landscapes. Through meticulous data analysis, they offer insights into evolutionary processes, adaptation, and speciation. Moreover, they underpin essential research in fields like pharmacogenomics and conservation genetics. In a scientific context, these tools are indispensable for the comprehensive exploration and comprehension of genetic diversity, thereby advancing our understanding of life's genetic intricacies.

References

Onda, Y., & Mochida, K. (2016). Exploring genetic diversity in plants using high-throughput sequencing techniques. *Current Genomics, 17*(4), 358-367.

Marsh, J. I., Hu, H., Gill, M., Batley, J., & Edwards, D. (2021). Crop breeding for a changing climate: Integrating phenomics and genomics with bioinformatics. *Theoretical and Applied Genetics, 134*, 1677-1690.

M Perez-de-Castro, A., Vilanova, S., Cañizares, J., Pascual, L., M Blanca, J., J Diez, M., ... & Picó, B. (2012). Application of genomic tools in plant breeding. *Current genomics, 13*(3), 179-195.

Tripodi, P. (2022). Next generation sequencing technologies to explore the diversity of germplasm resources: Achievements and trends in tomato. *Computational and Structural Biotechnology Journal.*

Della Coletta, R., Qiu, Y., Ou, S., Hufford, M. B., & Hirsch, C. N. (2021). How the pan-genome is changing crop genomics and improvement. *Genome biology, 22*(1), 1-19.

Zhang, H., Mittal, N., Leamy, L. J., Barazani, O., & Song, B. H. (2017). Back into the wild—Apply untapped genetic diversity of wild relatives for crop improvement. *Evolutionary Applications, 10*(1), 5-24.

Mahmood, U., Li, X., Fan, Y., Chang, W., Niu, Y., Li, J., ... & Lu, K. (2022). Multi-omics revolution to promote plant breeding efficiency. *Frontiers in Plant Science, 13*, 1062952.

Voss Fels, K., & Snowdon, R. J. (2016). Understanding and utilizing crop genome diversity via high resolution genotyping. *Plant biotechnology journal, 14*(4), 1086-1094.

Bohra, A. (2013). Emerging paradigms in genomics-based crop improvement. *The Scientific World Journal, 2013.*

Varshney, R. K., Bohra, A., Yu, J., Graner, A., Zhang, Q., & Sorrells, M. E. (2021). Designing future crops: genomics-assisted breeding comes of age. *Trends in Plant Science, 26*(6), 631-649.

Alqudah, A. M., Sallam, A., Baenziger, P. S., & Börner, A. (2020). GWAS: fast-forwarding gene identification and characterization in temperate cereals: lessons from barley–a review. *Journal of advanced research, 22*, 119-135.

Xiao, Q., Bai, X., Zhang, C., & He, Y. (2022). Advanced high-throughput plant phenotyping techniques for genome-wide association studies: A review. *Journal of advanced research, 35*, 215-230.

Jenkins, S., & Gibson, N. (2002). High-throughput SNP genotyping. *Comparative and Functional Genomics, 3*(1), 57-66.

Fernie, A. R., & Schauer, N. (2009). Metabolomics-assisted breeding: a viable option for crop improvement?. *Trends in genetics, 25*(1), 39-48.

Scossa, F., Alseekh, S., & Fernie, A. R. (2021). Integrating multi-omics data for crop improvement. *Journal of plant physiology, 257*, 153352.

Zhang, J., Yang, J., Zhang, L., Luo, J., Zhao, H., Zhang, J., & Wen, C. (2020). A new SNP genotyping technology Target SNP-seq and its application in genetic analysis of cucumber varieties. *Scientific reports, 10*(1), 5623.

lseekh, S., Kostova, D., Bulut, M., & Fernie, A. R. (2021). Genome-wide association studies: assessing trait characteristics in model and crop plants. *Cellular and Molecular Life Sciences, 78*, 5743-5754.

Akbari, M., Wenzl, P., Caig, V., Carling, J., Xia, L., Yang, S., ... & Kilian, A. (2006). Diversity arrays technology (DArT) for high-throughput profiling of the hexaploid wheat genome. *Theoretical and applied genetics, 113*, 1409-1420.

Sansaloni, C., Petroli, C., Jaccoud, D., Carling, J., Detering, F., Grattapaglia, D., & Kilian, A. (2011, December). Diversity Arrays Technology (DArT) and next-generation sequencing combined: genome-wide, high throughput, highly informative genotyping for molecular breeding of Eucalyptus. In *BMC proceedings* (Vol. 5, No. 7, pp. 1-2). BioMed Central.

Chapter-9

Genomic Approaches in Improving Heterosis

Navya Ravuri [1]**, Deepa Dharshini** [1]
and Talari Akshay Kumar [2]

1 - Tamil Nadu Agricultural University, Coimbatore.

2 – Agricultural College, Jagityal, PJTSAU.

9.1. Introduction

Any breeding strategy aims to achieve a higher level of performance with respect to yield and other characteristics in the hybrid or a progeny compared to both the parents. This ability of a hybrid or a progeny to surpass the genetic performance of both parental lines or the mean performance of the parents is named as Heterosis by Shull, while studying the maize hybrids. Heterosis also commonly known as hybrid vigor, which describes the progeny vigor compared to that of the parents. It's discovery and understanding have evolved over decades, with significant contributions from various scientists. Many studies conducted through the past century failed to describe the exact genetic reasons underlying the phenomenon of heterosis in plants. However, different hypotheses were given by many scientists explaining the molecular basis of this phenomenon like dominance hypothesis by Davenport (1910), Epistasis hypothesis by Shull, Over-dominance hypothesis by East and Shull (1903) each having a few limitations.

Over years, hybrid breeding has been one of the efficient ways to enhance the yield capacity in various crops. Hybrids have a great ability to withstand the present-day climate change scenario and biotic stresses, when compared to their parental lines. Now, hybrids are available in almost all the cultivated crops. So, understanding the mechanism underlying heterosis and utilizing it for development of hybrids is an easy and efficient way to improve crop yields. With the recent developments in molecular biotechnology and new genomic approaches available, various research attempts have been made to understand the potential genetic or molecular basis of heterosis. In this chapter, we will discuss briefly about the modern genetic approaches for understanding and enhancing heterosis in plants.

9.2. Modern Approaches for Improving Heterosisin Crop Plants

9.2.1. Trasncriptomics and Proteomic Studies

Phenotype is the result of genetic information from various intra-cellular processes like DNA transcription, RNA translation and metabolic processes. Heterosis is observed mainly due to uncontrollable gene combinations in various plant tissues from both the parental lines. These genes may be responsible for many physiological and biochemical processes in plants like signal-transduction, cell-division, transcription and translation activities, gene interactions in defense or in stress response, metabolic processes *etc.* So, studying the combining ability of such genes can partly help to understand the mechanism of heterosis for achieving crop improvement. The available modern molecular biotechnology tools like DNA microarray studies, QTL mapping and high-throughput sequencing approaches provide an easy way to determine the differences in gene expressions between various genotypes.

Transcriptomic studies can measure the relative contribution of each allele of a gene to the hybrid vigor. Studies on relationship between heterosis and transcriptomics has proved that the additive and non-additive gene interactions might be the main reasons for these gene-expression differences in hybrids and their parental genotypes.

Proteomic study is another approach to determine the molecular basis of heterosis. Proteomics is defined as a detailed study of the structure and function of different proteins. Proteomic studies on hybrids and parental lines revealed that the end expression differences are mainly due to non-additive gene interactions. Proteomics acts as an effective supplement to the transcriptomic studies. Recent studies proved that there is a difference in the expression of stress response proteins and also carbon metabolism in the heterotic maize hybrids when compared to non-heterotic hybrids. Studies on rice hybrids proved that differences in the expression of transcription factors and their associated promoter elements are two main factors effecting the heterosis.

9.2.2. Epigenetic Studies

Epigenetics is an advance area of study, that involves the phenotypic differences due to altered or modified gene functions, but not the differences in DNA sequences. It can be due to histone modifications, non-coding RNAs or methylation of DNA *etc.* These factors may also play a key role in determining the hybrid performance.

- ❏ **DNA Methylation:** The DNA methylation levels varies between parents and their progeny. Studies on *Arabidiopsis* hybrids have shown that the reduction in heterosis is due to the reduction in the DNA methylation levels of few contributing genes and also due to the knockdown METHYLTRANSFERASE-1 gene effect.

- ❏ **Histone Modification:** Histone modifications like acetylation and DNA methylation can affect the regulation of variations in gene expression. In *Arabidiopsis,* heterosis is associated with differences in histone modification in the promoter regions of genes like CCA1 and LHY. During the pathogen attack, the CCA1 gene expression is highly regulated by a definite histone modification at various time points of a day.

- ❏ **Small RNAs:** Small RNA like small interfering RNAs (siRNAs), microRNA (miRNAs) and transactivating RNAs also have key role in epigenetics. Differential expression of these siRNAs and their heterotic effects on different target genes may thereby affect the grain yield of hybrids to a great extent. Shen *et al.,* 2010 discovered *heterosis* being regulated by non-additive gene effects in *Arabidiopsis.* In different hybrids of rice, the differences in gene expression are regulated by miRNAs. Sequencing of RNA library of 21 maize inbred lines revealed that, heterosis of grain yield is negatively correlated to the variations in the siRNA gene expressions in the parental lines.

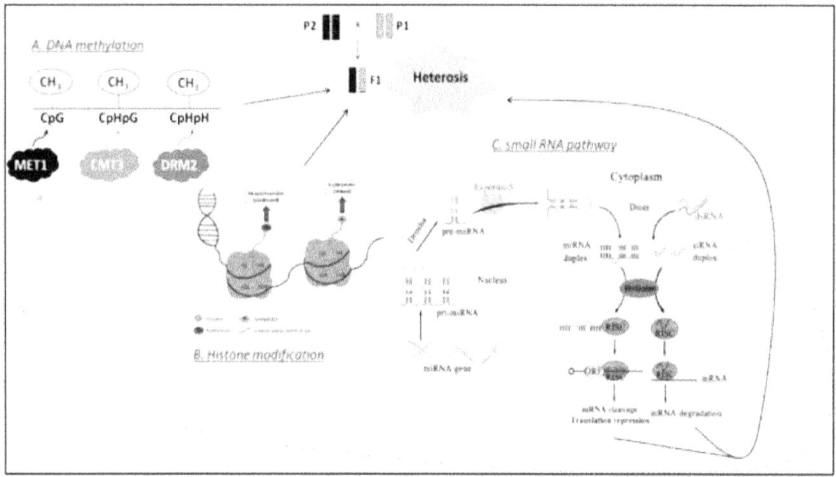

The above figure summarizes the different epigenetic approaches for studying heterosis in plants. A) DNA methylation occurs due to the addition of methyl group (CH_3) to the 5′ end of the cytosine base in the DNA sequence. B) Histone modification refers to the changes in the histone proteins that can significantly affect the associated DNA sequence, thereby affecting the gene expression. C) siRNAs can directly affect the gene expression by silencing the expression of a gene, by gene editing through RNA-induced gene complex (RISC).

9.3. Phenomics Approach

Various heterotic genes responsible for raising crop yields are now being hunted down using advanced genomic approaches, particularly transcriptomics, but with contrary results. Because heterotic effects of a gene are highly influenced by environment, genomic analysis alone will not provide sufficient results. So, future research studies should focus on an integrated genomics approach of comprehensive quantitative trait locus (QTL)-based phenotyping, followed by map-based cloning of such QTLs. The phenomics approach identifies gene loci controlling heterotic phenotypes, and helps in understanding of the role of heterosis in evolution and the domestication of various crop plants.

One among the most promising approaches at the molecular level was emerged through the availability of molecular or DNA markers, that was used to identify the genomic regions that contribute to heterosis for a trait of interest. Specific genes/QTL for individual traits contributing to heterosis for desirable traits can be used to enhance the performance of hybrids by transferring them into parental inbred lines through MAS but may be very challenging. The complex trait 'heterosis' is expected to be reflected by many genes, their wide genomic distribution, the combination and interaction of which may depend on the organism and trait under study (Korn *et al.*, 2008; Li *et al.*, 2008). To exploit heterosis by its best it is much needed to understand the nature of dominance, epistatic properties of these genes and how they interact with the environment (Coors and Pandey, 1999).

Also, marker-based QTL studies are inherently inefficient at detecting epistasis and one cannot exclude even the slightest possibility that some level of epistasis is occurring. Mapping and cloning of QTLs with heterotic gene effects requires more rigorous approaches, particularly considering the global phenotyping that is much expensive and time consuming.

An alternative phenomics platform for each crop was proposed which would include a database of unbiased measurement of multiple traits (*e.g.,* each component of total yield is treated as individual trait and all such traits are recorded in well-characterized environmental conditions in term of seasons, locations, and years (Lippman and Zamir, 2007).

9.4. eQTL Analysis

The variation observed in the level of gene expression as a consequence of genotypic differences is termed as an expression level polymorphism (ELP), and the QTL responsible for such type of variation has been described as eQTL (Jansen and Nap, 2001; Doerge 2002; Gibson and Weir, 2005). Advances in QTL mapping or analysis and genetic genomics involving identification of expression QTL (eQTL), have led to significant progress in genetic differentiation of complex traits likely heterosis.

When the abundance of transcript is treated as a continuous trait for the mapping purpose, it is referred as an expression trait (eTrait). The eQTL analysis, when compared to classical quantitative trait analysis, may provide relatively more detailed information about a gene network controlling a trait, because in this analysis, data on thousands of gene expression traits are recorded simultaneously, also there is a one-to-one association between an eTrait and a gene with respect to the expression profile data obtained from the mapping population. Provided with these tools, expression quantitative trait locus (eQTL) analysis has been applied to study inheritance of thousands of similar quantitative traits in the hope to find the basis of genetic control of their transcriptional regulations (Schadt *et al.*, 2003).

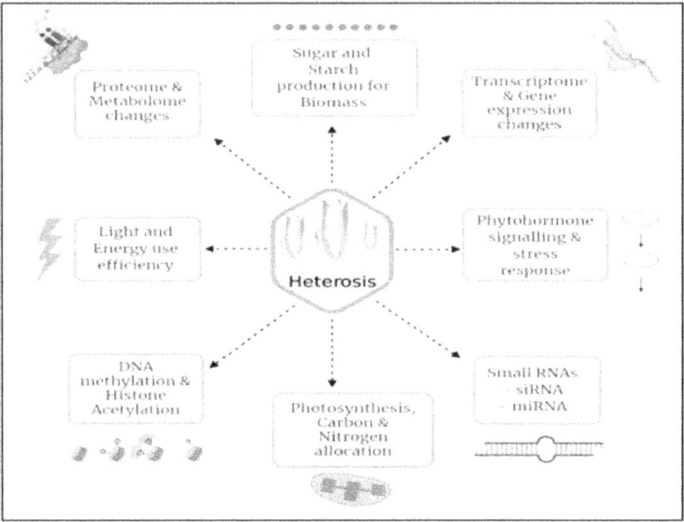

Different molecular mechanisms and physiological pathways that can explain heterosis are summarized in the above picture.

The above figure summarizes the future perspectives for developing the breeding strategies. Once the superior allele combinations are identified using GWAS or linkage analysis studies and understanding the molecular pathways underlying the heterotic effects of such genes, the potential hybrid combinations for a higher heterosis performance can be selected easily. Superior hybrids of such combinations can be achieved by genome-editing or gene-pyramiding technologies. The hybrids thus developed can be maintained by asexual propagation through seeds.

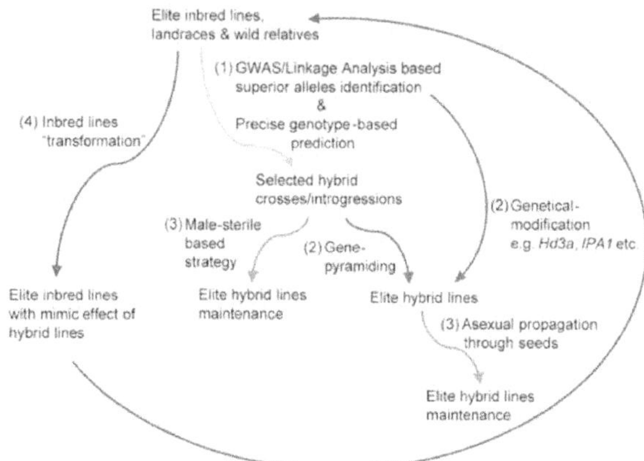

9.5. Future Perspectives

Hybrid breeding has been proven to be one of the successful and reliable ways to improve the yield potential in many crop species. But, characterizing the genetic reasons behind heterosis has been a difficult task for decades. The main reason underlying this failure is considered to be multiple genes controlling the heterosis in plants. Though there are various advanced genomic approaches to study the mechanisms underlying heterosis, no single approach discussed in this chapter has been completely successful to explain the potential genetic reasons for heterosis. Therefore, a potential future-research should involve a multi-omics approach combining the high-throughput tools, which can be effective, as it can provide a real-time analysis of multiple genes and proteins simultaneously at various developmental stages of a plant. A better knowledge on heterotic effects of genes in plants can help in better designing of breeding schemes for developing the elite hybrids with target characters in a single hybrid.

References

Liu, J., Li, M., Zhang, Q., Wei, X and Huang, X. Exploring the molecular basis of heterosis. *Journal of Integrative Plant Biology*. 2019; 62(3): 287-29.

Pansare, M., Walia, P., Patil, K and Gopera, S. Unravelling the Genetic Enigma: Exploring the Molecular Basis of Heterosis. *International Journal of Environment and Climate Change*. 2023;13(9): 228-237.

Rehnman AU et al., 2021. Revisisting plant heterosis- from field scale to molecules. *Genes (Basel)*. 24; 12(11):1688.

Sanghera, G. S., Wani, S. H., Hussain, W., Shafi, W., Haribhushan, A and Singh, N. B. The magic of heterosis: new tools and complexities. *Nature Science*. 2011; 9(11): 42-53.

Wu, X., Liu, Y., Zhang, Y and Gu, R. (2021). Advances in research on the mechanism of heterosis in plants. *Frontiers in Plant Science*. 2021; Vol. 12.

Chapter-10

Breeding for Abiotic Stress Tolerance Through Plant Genomics Approach

Pravin Kumar K*, Sathish Kumar R, Vinoth Kumar G, Arun Kumar M

Ph. D Scholar, Centre for Plant Breeding and Genetics, TNAU, Coimbatore
Corresponding author: pravinagri378@gmail.com

10.1. Introduction

Abiotic stresses, such as salinity, drought, extreme temperatures, and heavy metal exposure, pose significant challenges to agriculture, potentially leading to crop yield losses of up to 50%. These stressors disrupt the fundamental physiological and biochemical processes in plants, causing disturbances in cellular water and ion balance. In response to these stressors, hundreds of genes and their products are involved at the transcriptional and translational levels, contributing to the complex mechanisms of stress tolerance. Among these abiotic stresses, drought is particularly widespread and devastating factor in agriculture, which prompting extensive research to dissect its components and understand the mechanisms of plant tolerance. However, the intricate interplay of genetic and environmental factors in abiotic stress response often complicates the traditional breeding strategies. To address these challenges and meet the demands of a growing global population, agriculture has turned to innovative approaches empowered by recent technological advancements. High-throughput tools aimed at exploring and harnessing plant genomes have emerged as powerful solutions. These genomics-based approaches delve into the entire genome, including both genic and intergenic regions, providing deep insights into how plants molecularly respond to stress. These insights, in turn, guide the development of specific strategies to enhance crop resilience.

- ❏ Functional genomics,
- ❏ Structural genomics, and
- ❏ Comparative genomics are three broad classes of genomics approaches that play crucial roles in improving crops' ability to withstand abiotic stresses.

These approaches, while distinct, often overlap in methodologies and outcomes, collectively contributing to a comprehensive understanding of plant stress responses. In this context, this chapter will explore and categorize these genomics approaches, highlighting their significance in unraveling signaling cascades, regulatory mechanisms, and the components of stress tolerance mechanisms. Additionally, we will discuss how these approaches provide critical insights into the role of signaling networks and transcriptional regulators in controlling gene expression in response to abiotic stressors.

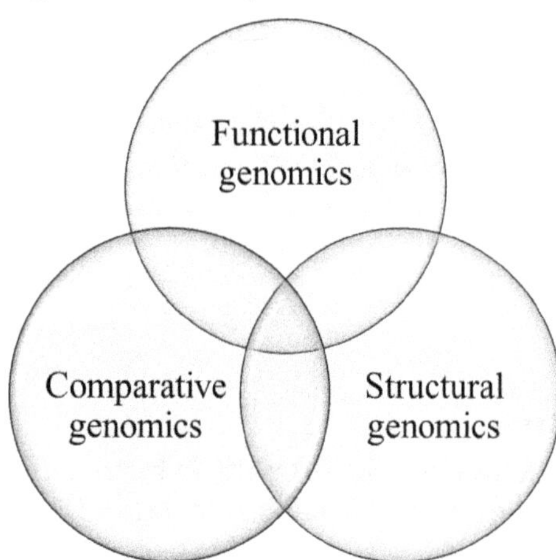

Fig.10.1 Genomic Approaches for Developing Tolerance in Abiotic Stresses

10.2. Functional Genomic Approaches Role in Elucidating Abiotic Stress Tolerance

Functional genomic approaches play a pivotal role in unraveling the mechanisms underlying abiotic stress tolerance in plants, particularly in response to osmotic stress such as drought and salinity. These approaches are instrumental in understanding how plants respond and adapt to environmental stressors by modulating the expression of numerous genes, which subsequently impacts cellular, physiological, and biochemical processes. Transcriptome analysis, facilitated by microarray technology, provides a potent platform for studying global gene expression changes during osmotic stress. This comprehensive analysis allows researchers to gain insights into the intricate molecular, metabolic, physiological, cellular, and morphological challenges that arise in response to stress. Several key aspects of functional genomics have significantly contributed to advancing our understanding of abiotic stress tolerance mechanisms:

(a) **Gene Discovery** - Functional genomics techniques help to identify stress-responsive genes. By analyzing gene expression profiles under stress conditions, researchers can pinpoint the specific genes that are activated or repressed in response to stress. By this way, researchers can pick-out the novel genes for the particular stressor. This gene discovery is essential for understanding how plants perceive and respond to stress. To identify abiotic stress-related genes, genomic approaches involve constructing an Expressed Sequence Tags (ESTs) from cDNA libraries of stressed plant tissues. These libraries can be used a reference for comparing the libraries developed from the various stress treatments and at different development stages of plant tissues. Detailed information about these libraries and the number of ESTs they produce is available at the National Center for Biotechnology Information (NCBI) dbEST. To enrich these datasets with stress-responsive genes, specialized sequencing programs are essential. They create cDNA libraries from stress-treated plant tissues and organs across different plant species and developmental stages. Since these datasets are usually from non-normalized libraries, they provide insights into the relative expression levels of stress-responsive genes. Cluster analysis of EST sequences from these libraries reveals the number of genes, their content, and potential gene families involved in stress responses. By comparing these sequences with the Swissprot database using BLASTX, putative functions are assigned to stress-responsive genes. This approach yields a valuable resource, offering insights into stress-responsive genes across various species. Recent efforts have been identified abundantly expressed ESTs in libraries from stress-treated plants, such as the halophyte *Thelungiella halophila* and monocot species like barley, wheat, maize, and rice (Wang *et al.*, 2004, sreenivasulu *et al.*, 2006). These datasets help identify stress-regulated genes and shed light on the underlying regulatory and metabolic networks involved in plant stress responses.

(b) **Transcriptome Analysis** - Transcriptome analysis, often performed using technologies like microarrays or RNA sequencing (RNA-seq), allows for a comprehensive examination of gene expression changes during abiotic stress. This provides insights into the candidate genes and pathways involved in stress responses. In contrast to digital *in silico* quantification of gene expression levels based on EST counts, various experimental approaches offer high-throughput assessment of thousands of genes in control and stress-treated tissues at different developmental stages. These approaches provide valuable insights into gene expression patterns and their functions, particularly in the context of stress tolerance.

- ❒ **SAGE (Serial Analysis of Gene Expression)** - SAGE is a technique that allows the quantification of gene expression by sequencing short tags from cDNA libraries. It enables researchers to evaluate the expression of a large number of genes simultaneously in both control and stressed plant tissues.

- ❒ **Array-Based Transcript Profiling Technologies** - DNA microarrays and other array-based platforms offer the ability to simultaneously measure the expression of thousands of genes. Researchers can use these platforms to compare gene expression profiles between control and stressed plant tissues, uncovering stress-responsive genes and pathways.

- ❒ **Quantitative Real-Time PCR (qRT-PCR)** - qRT-PCR is a precise and widely used technique for quantifying gene expression. It allows for the quantification of transcripts of interest, making it a valuable tool for validating gene expression data obtained from other high-throughput methods. Several studies have applied these transcriptome profiling approaches to model species such as *Arabidopsis* and rice. These studies have revealed numerous stress-related pathways beyond the well-described stress-related genes. This research has expanded our understanding of how plants respond to abiotic stress at the molecular level, shedding light on novel genes and mechanisms involved in stress tolerance.

(c) Pathway and Network Analysis - Functional genomics enables the investigation of entire gene networks and pathways involved in stress tolerance. This holistic approach helps identify key regulatory genes and interactions critical for plant adaptation to stress. Recent research has placed significant emphasis on understanding the mechanisms behind dehydration responses in plant vegetative tissues, particularly in the context of gene expression linked to desiccation tolerance. These responses are often triggered by the plant hormone abscisic acid (ABA) and involve specific transcription factors such as ABA-responsive element binding factors (ABF), MYC, and MYB. Additionally, ABA-independent pathways, mediated by drought-responsive element binding factors (DREB), have been identified (Akpinar, B.A *et al.*, 2013). Transcriptome studies revealed that ABA plays a chief role not only in drought-specific responses but also in the response to cold and salinity stresses. This specifies a cross-talk between different stress response pathways (Seki *et al.*, 2002). In a specific laboratory effort, the focus has been on identifying regulatory genes involved in the desiccation tolerance program during the natural development of seed embryos. This program is a crucial part of the maturation process, during which seed embryos become desiccation tolerant. Signaling to activate the desiccation tolerance program in developing seed embryos occurs through the action of abscisic acid. ABA activates genes involved in both ABA biosynthesis and signaling pathways, enabling the development of desiccation tolerance (Sreenivasulu *et al.*, 2007). Understanding these regulatory mechanisms is essential for improving crop resilience to dehydration and other abiotic stresses, ultimately contributing to enhanced agricultural productivity and stress tolerance in plants.

(d) Functional Characterization - Functional genomics techniques, including gene knockout or knockin, suppression or overexpression, can be used to evaluate the roles of specific genes in stress tolerance. By manipulating the expression of target genes, researchers can determine their impact on stress responses and plant performance. Stress perception and the intricate signaling pathways that follow are crucial for plants to endure harsh environmental conditions like drought, salt, and extreme temperatures. Osmotic stress and oxidative stress often intersect, sharing common components in their signaling networks. Numerous proteins, including histidine kinases, receptor-like kinases, MAPK cascades, and calcium-dependent protein kinases, have been implicated in the sensing and transmission of stress signals within plants. For instance, histidine kinase AtHK1 in *Arabidopsis* appears to serve as a potential sensor for stress, as it is upregulated during salt and low-temperature stress. Additionally, receptor-like protein kinase NtC7, induced under abiotic stress, confers osmotic stress tolerance in transgenic tobacco plants. Mitogen-activated protein kinase (MAPK) cascades act as convergence points for cross-talk between different stress responses. Overexpressing *Arabidopsis* MAPK kinase 2 (MKK2) leads to increased activity of downstream MPK4 and MPK6, resulting in enhanced stress tolerance. Calcium-dependent protein kinases (CDPKs) also play roles in osmotic signaling, as demonstrated by the overexpression of rice CDPK7, which triggers stress-responsive genes and enhances tolerance to salt, drought, and cold stresses (Sreenivasulu *et al.*, 2006). However, the full understanding of these abiotic stress signaling pathways remains fragmented, and further systematic dissection is needed through forward and reverse genetics approaches.

(e) Identification of Regulatory Elements - Functional genomics can display the presence of regulatory elements, such as transcription factors and microRNAs, that govern gene expression under stress conditions. Understanding how these elements modulate gene expression is essential for engineering stress-tolerant plants. One important facet of current research using functional genomic tools is the identification of key regulators based on gene expression patterns in response to multiple stress interactions. Genome-wide transcriptome analysis has unveiled hundreds of genes encoding transcription factors that are either induced or repressed by various environmental stresses (Chen and Zhu, 2004). These transcription factors exhibit intricate and complex expression patterns, suggesting that stress tolerance and resistance are finely controlled at the transcriptional level within a highly complex gene regulatory network.

Chen *et al.* (2002) classified these transcription factors into two groups:

- ❏ those specifically regulated by abiotic stress (class I) and
- ❏ those influenced by both biotic and abiotic stresses (class II) in *Arabidopsis*.

Class I included around 20 genes preferentially induced by abiotic stresses like salinity, osmotic stress, cold, and jasmonic acid treatments. These transcription factors comprised DRE/CRT binding factors activated by cold stress, CCA1 and Athb-8 (hormone-regulated), Myb proteins, bZIP/HD-ZIPs, and AP2/EREBP domain proteins. Seki *et al.* (2002) employed a full-length cDNA microarray with 7000 *Arabidopsis* cDNAs to identify target genes induced by cold, drought, and salinity, along with stress-related transcription factor families such as DREB, ERF, WRKY, MYB, bZIP, helix-loop-helix, and NAC. These findings suggest that significant cross-talk between salt and drought stress signaling processes compared to salt and cold stresses. During short-term cold acclimation transcriptome studies in *Arabidopsis*, Fowler and Thomashow (2002) noted rapid expression of CBF1 (DREB1b), CBF2 (DREB1c), and CBF3 transcripts (DREB1a). Long-term up-regulated genes included transcription factors like putative zinc finger protein, R2R3-Myb transcription factor AtMYB73, H-protein promoter binding factor 2a, HD-Zip protein AthB-12, and two AP2 domain proteins, RAP2.7 and RAP2.1. In a similar vein, transcriptome responses to dehydration, salinity, and ABA in sorghum seedlings identified around 22 transcription factors, including ABF (bZIP factors), DREB (AP2/EREBP family), HD-ZIP, and MYB factors. These regulators have also been recognized as stress-responsive in other model species like *Arabidopsis* and rice. However, there is a pressing need to confirm the roles these transcription factors play in stress response networks, a vital step in designing plants capable of withstanding various environmental stresses.

(f) Genetic Mapping - Functional genomics can be integrated with genetic mapping approaches to identify quantitative trait loci (QTLs) associated with abiotic stress tolerance. These genomic regions contain genes responsible for stress resilience and can be targeted in breeding programs. Conventional breeding programs, while successful in developing abiotic stress-tolerant plants in some cases, often lead to a trade-off between stress tolerance and yield. The emergence of genomic resources has opened two promising avenues for enhancing stress tolerance in crops. Firstly, functional genomic approaches enable the identification of stress-tolerant genes, which can then be introduced into target crops. Secondly, quantitative trait loci (QTLs)/genes associated with stress tolerance in germplasm collections can be identified, developed molecular markers which can be implemented in marker-assisted breeding programs. Although functional genomics has provided insights into abiotic stress responses, translating these findings to field conditions remains challenging. Integrating functional genomics knowledge into practical breeding programs through genomics-assisted breeding is crucial. It involves exploring variations in abiotic stress tolerance between species and accessions, introgressing tolerant wild species into elite genomes, and utilizing -omics technologies to understand regulatory mechanisms governing yield under stress conditions. Wild species and landraces offer valuable genetic diversity and resistance alleles which is not present in cultivated gene pools. For instance, wild barley exhibits significant variation and tolerance to salinity. Similarly, wild species in the *Triticeae* tribe have shown discrimination in K^+/Na^+ uptake. These resources can be tapped to target genetic loci controlling abiotic stress in cultivated species. Breeding programs can employ approaches like Advanced Backcross QTL Analysis, where wild species are backcrossed with elite cultivars to simultaneously discover and transfer desirable QTLs. Marker-assisted selection aids in introgressing QTL alleles for drought tolerance, with genomic approaches allowing the tracking of QTL effects at a single-gene level.

"Genetical Genomics" is an emerging approach combining gene expression profiling and marker-based fingerprinting in segregating populations to elucidate genetic networks underlying traits. It can be applied to identify sets of trait-related genes and pathways controlling tolerance and yield. For example, introgression lines, incorporating an alien genome conferring stress tolerance into an elite background, are used to track characteristics for both yield and tolerance. Expression profiling from these lines helps detect expression QTLs (eQTLs) and develop trait-linked molecular markers, enhancing molecular breeding strategies. While the "genetical genomics" approach is still evolving, it holds promise for accelerating success in improving agronomic traits, including abiotic stress tolerance, by integrating genomic insights with genetic-based breeding efforts.

(g) Systems Biology - Functional genomics contributes to systems biology, allowing researchers to model and simulate the complex interactions between genes, proteins, and metabolites involved in stress responses. This systems-level understanding helps in predicting plant behavior under stress conditions. Systems biology plays a crucial role in the development of abiotic stress tolerance in plants through functional genomics. It provides a comprehensive and integrated approach to understanding the complex molecular and cellular mechanisms involved in a plant's response to environmental stresses such as drought, salinity, extreme temperatures, and heavy metals (Cramer, G. R., *et al.*, 2011). Here's how systems biology contributes to enhancing abiotic stress tolerance in plants:

  ❐ **Holistic Understanding** - Systems biology allows researchers to view plant responses to abiotic stress as a whole, considering all the interconnected components, including genes, proteins, metabolites, and signaling pathways. This holistic understanding is essential for identifying key regulators and targets for genetic and molecular manipulation.

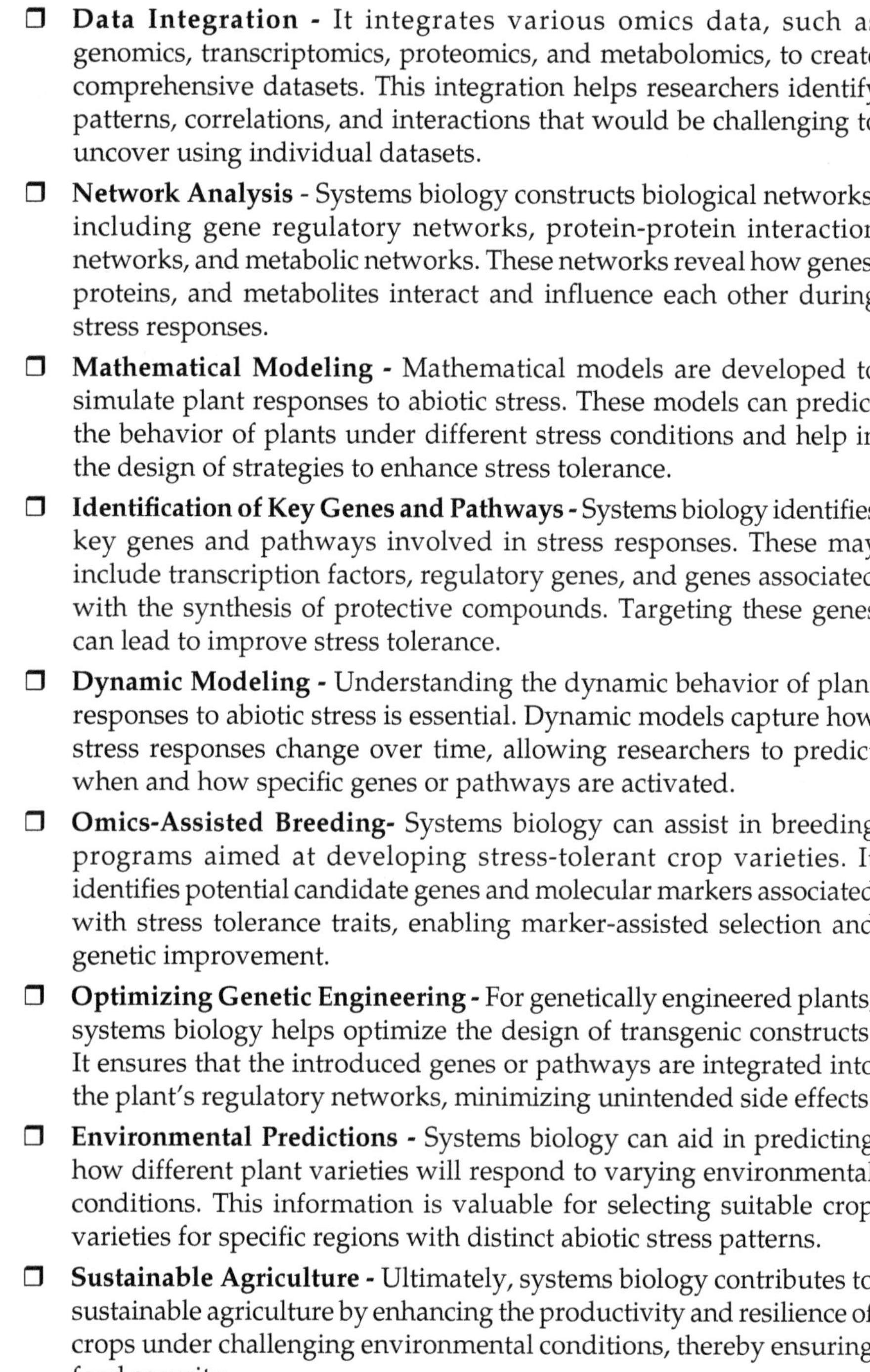

- ❏ **Data Integration -** It integrates various omics data, such as genomics, transcriptomics, proteomics, and metabolomics, to create comprehensive datasets. This integration helps researchers identify patterns, correlations, and interactions that would be challenging to uncover using individual datasets.

- ❏ **Network Analysis -** Systems biology constructs biological networks, including gene regulatory networks, protein-protein interaction networks, and metabolic networks. These networks reveal how genes, proteins, and metabolites interact and influence each other during stress responses.

- ❏ **Mathematical Modeling -** Mathematical models are developed to simulate plant responses to abiotic stress. These models can predict the behavior of plants under different stress conditions and help in the design of strategies to enhance stress tolerance.

- ❏ **Identification of Key Genes and Pathways -** Systems biology identifies key genes and pathways involved in stress responses. These may include transcription factors, regulatory genes, and genes associated with the synthesis of protective compounds. Targeting these genes can lead to improve stress tolerance.

- ❏ **Dynamic Modeling -** Understanding the dynamic behavior of plant responses to abiotic stress is essential. Dynamic models capture how stress responses change over time, allowing researchers to predict when and how specific genes or pathways are activated.

- ❏ **Omics-Assisted Breeding-** Systems biology can assist in breeding programs aimed at developing stress-tolerant crop varieties. It identifies potential candidate genes and molecular markers associated with stress tolerance traits, enabling marker-assisted selection and genetic improvement.

- ❏ **Optimizing Genetic Engineering -** For genetically engineered plants, systems biology helps optimize the design of transgenic constructs. It ensures that the introduced genes or pathways are integrated into the plant's regulatory networks, minimizing unintended side effects.

- ❏ **Environmental Predictions -** Systems biology can aid in predicting how different plant varieties will respond to varying environmental conditions. This information is valuable for selecting suitable crop varieties for specific regions with distinct abiotic stress patterns.

- ❏ **Sustainable Agriculture -** Ultimately, systems biology contributes to sustainable agriculture by enhancing the productivity and resilience of crops under challenging environmental conditions, thereby ensuring food security.

Systems biology in functional genomics provides a powerful toolbox for deciphering the molecular mechanisms underlying abiotic stress tolerance in plants. It offers a systems-level perspective that is essential for designing effective strategies to develop stress-tolerant crop varieties, contributing to global efforts to address food security challenges in a changing climate. functional genomics plays a pivotal role in advancing our knowledge of abiotic stress tolerance in plants. It helps identify key genes, pathways, and regulatory mechanisms that enable plants to adapt and thrive in challenging environments. This knowledge is invaluable for the development of stress-tolerant crop varieties, which is crucial for ensuring food security in the face of changing climate conditions and increasing global demand for agricultural products.

10.3. Functional Genomics in Crop Improvement

☐ **Gene Discovery** - Functional genomics helps to detect genes responsible for specific traits, such as disease resistance, drought tolerance, and yield enhancement, by studying gene expression patterns and functions.

☐ **Gene Editing** - CRISPR-Cas9 and other gene-editing technologies, informed by functional genomics data, enable precise modification of crop genomes to introduce or enhance desirable traits.

☐ **Targeted Breeding** - Functional genomics provides insights into the genetic basis of traits, allowing breeders to select plants with desired genetic markers for more efficient traditional breeding programs.

☐ **Validation of Candidate Genes** - Functional genomics experiments validate the role of candidate genes in trait improvement, increasing confidence in their use for crop enhancement.

☐ **Pathway Analysis** - Understanding gene pathways through functional genomics helps in identifying the key genes and regulatory networks involved in important physiological processes, aiding crop modification.

10.4. Structural Genomics Approaches Role In Elucidating Abiotic Stress Tolerance

Structural genomics, on the other hand, focuses on the physical structure of genomes, including the identification, location, and order of genomic features along chromosomes. This field contributes to the development of physical maps and molecular markers. Key aspects of structural genomics (Varshney, R. K., *et al.*, 2005)

10.4.1. Genome Sequencing and Mapping

Advances in DNA sequencing technologies, such as next-generation sequencing (NGS), have enabled the generation of whole genome sequences. These sequences provide detailed information about gene content, regulatory regions, and repetitive elements. However, sequencing large and complex genomes, such as those of wheat and barley, remains challenging.

10.4.2. Physical Mapping

Physical mapping involves creating chromosome-specific maps using techniques like Bacterial Artificial Chromosome (BAC) libraries. These maps help assemble genomes and locate genes of interest.

10.4.3. Molecular Markers

Molecular markers, particularly Single Nucleotide Polymorphisms (SNPs), are widely used in genomics. SNPs are identified through genome or transcriptome resequencing and are valuable for marker-assisted selection (MAS) in breeding programs.

10.4.4. Map-Based Cloning (MBC)

MBC is used to isolate genes or quantitative trait loci (QTLs) linked to specific traits. It relies on high-density physical maps to identify and clone genes of interest.

10.4.5. Genome Sequencing

Structural genomics techniques like next-generation sequencing (NGS) provide complete or draft genome sequences, enabling researchers to access comprehensive genetic information for crop species.

10.4.6. Genome Assembly

Structural genomics facilitates the assembly of complex genomes, including those of polyploid crops, providing insights into genome structure and organization.

10.4.7. Identification of Repetitive Elements

Structural genomics helps identify repetitive elements, transposons, and mobile genetic elements that impact genome stability and evolution.

10.4.8. Comparative Genomics

Structural genomics data form the basis for comparative genomics, allowing researchers to compare crop genomes with related species and identify conserved regions and functional elements.

10.5. Comparative Genomics Approaches Role in elucidating Abiotic Stress Tolerance

Comparative genomics is a field of genomics that focuses on comparing the genomes of different species to uncover similarities, differences, and evolutionary relationships. It involves the analysis of genomic data from multiple organisms to gain insights into various aspects of genome evolution, gene function, and adaptation. Here are key aspects of comparative genomics (Morrell, P. L., *et al.*, 2012)

10.5.1. Genome Evolution

Comparative genomics helps researchers understand how genomes have evolved over time. By comparing the genomes of closely related species or those from different lineages, scientists can trace the evolutionary history of genes, regulatory elements, and entire genomic regions.

10.5.2. Identification of Conserved Genes

Through comparative genomics, conserved genes that are present across multiple species can be identified. These genes often encode essential functions and are fundamental to the biology of diverse organisms.

10.5.3. Orthologous and Paralogous Genes

Comparative genomics distinguishes between orthologous genes (genes in different species that evolved from a common ancestral gene) and paralogous genes (genes within the same species that evolved through gene duplication). Understanding these relationships provides insights into gene duplication events and functional diversification.

10.5.4. Gene Function Prediction

Comparative genomics can help predict the functions of genes in poorly studied organisms based on the functions of orthologous genes in well-characterized species. This is especially valuable when studying non-model organisms.

10.5.5. Gene Gain and Loss

By comparing genomes, researchers can identify genes that have been gained or lost during evolution. This can shed light on the genetic basis of specific adaptations or traits unique to certain species.

10.5.6. Structural Variation

Comparative genomics reveals structural variations, such as gene rearrangements, inversions, and deletions, that contribute to genome diversity. These variations can impact gene expression and phenotype.

10.5.7. Identification of Regulatory Elements

Comparative genomics aids in the discovery of regulatory elements, such as enhancers and promoters, that control gene expression. Conserved regulatory regions often indicate functional importance.

10.5.8. Evolutionary Relationships

Phylogenetic analysis, a component of comparative genomics, helps construct evolutionary trees that illustrate the relationships among species based on genomic data. This informs our understanding of species divergence and relatedness.

10.5.9. Functional Annotation

Comparative genomics contributes to the functional annotation of genomes by associating genes with known functions or pathways in other species. This is valuable for annotating genes in newly sequenced genomes.

10.6. Comparative Genomics in crop improvement

- ❐ **Trait Transfer** - Comparative genomics identifies genes in wild or related species that confer desirable traits. This information can be used to transfer these traits into crop varieties.

- ❐ **Evolutionary Insights-** Comparative genomics helps researchers understand the evolutionary history of crops, providing insights into their adaptation to changing environments.

- ❐ **Functional Annotation -** Comparative genomics aids in the functional annotation of crop genomes by comparing them to well-studied model organisms, helping to identify gene functions and regulatory elements.

- ❐ **Marker Development -** Comparative genomics identifies genetic markers, such as SNPs, that can be used for marker-assisted breeding to select for desired traits.

- ❐ **Disease Resistance -** By comparing genomes of resistant and susceptible varieties, comparative genomics identifies genes associated with disease resistance, guiding the development of disease-resistant crops.

- ❐ **Climate Resilience-** Comparative genomics helps uncover genetic variations and adaptations that confer resilience to environmental stresses like drought, salinity, and extreme temperatures.

- ❐ **Crop Wild Relatives -** Comparative genomics highlights the genetic diversity present in wild relatives of crops, offering a source of novel genes and traits for crop improvement.

10.7. Conclusion

These three genomics approaches collectively provide a wealth of genomic information that empowers crop scientists and breeders to develop more productive, resilient, and sustainable crop varieties to meet the world's growing food and agricultural challenges. By integration all three genomic approaches, the breeders (creators) can alter the genomic region, and modify the pathways involved in resistance in precise manner which paves way for developing a novel and climate resilience genotype for the adverse environment.

References

Akpınar, B. A., Lucas, S. J., & Budak, H. (2013). Genomics approaches for crop improvement against abiotic stress. *The Scientific World Journal, 2013.*

Chen, W., Provart, N. J., Glazebrook, J., Katagiri, F., Chang, H. S., Eulgem, T., & Zhu, T. (2002). Expression profile matrix of *Arabidopsis* transcription factor genes suggests their putative functions in response to environmental stresses. *The plant cell*, *14*(3), 559-574.

Chen, W. J., & Zhu, T. (2004). Networks of transcription factors with roles in environmental stress response. *Trends in plant science*, *9*(12), 591-596.

Fowler, S., & Thomashow, M. F. (2002). *Arabidopsis* transcriptome profiling indicates that multiple regulatory pathways are activated during cold acclimation in addition to the CBF cold response pathway. *The Plant Cell*, *14*(8), 1675-1690.Cramer, G. R., Urano, K., Delrot, S., Pezzotti, M., & Shinozaki, K. (2011). Effects of abiotic stress on plants: a systems biology perspective. *BMC plant biology*, *11*(1), 1-14.

Morrell, P. L., Buckler, E. S., & Ross-Ibarra, J. (2012). Crop genomics: advances and applications. *Nature Reviews Genetics*, *13*(2), 85-96.

Seki, M., Narusaka, M., Ishida, J., Nanjo, T., Fujita, M., Oono, Y., & Shinozaki, K. (2002). Monitoring the expression profiles of 7000 *Arabidopsis* genes under drought, cold and high-salinity stresses using a full-length cDNA microarray. *The Plant Journal*, *31*(3), 279-292.

Sreenivasulu, N., Radchuk, V., Strickert, M., Miersch, O., Weschke, W., & Wobus, U. (2006). Gene expression patterns reveal tissue-specific signaling networks controlling programmed cell death and ABA-regulated maturation in developing barley seeds. *The Plant Journal*, *47*(2), 310-327..

Sreenivasulu, N., Sopory, S. K., & Kishor, P. K. (2007). Deciphering the regulatory mechanisms of abiotic stress tolerance in plants by genomic approaches. *Gene*, *388*(1-2), 1-13.

Varshney, R. K., Graner, A., & Sorrells, M. E. (2005). Genomics-assisted breeding for crop improvement. *Trends in plant science*, *10*(12), 621-630.

Wang, H., Huang, Z., Chen, Q., Zhang, Z., Zhang, H., Wu, Y., & Huang, R. (2004). Ectopic overexpression of tomato JERF3 in tobacco activates downstream gene expression and enhances salt tolerance. *Plant molecular biology*, *55*, 183-192.

Chapter–11

Enhanced Photosynthetic Efficiency and Carbon Capture: Focus in Plant Breeding

Soorya. E[1] and Solaiyappan. M[2]

1   *Ph.D. Scholar, Department of Crop Physiology, Tamil Nadu Agricultural University*

2   *Ph.D. Scholar, Department of Crop Physiology, Indian Agricultural Research Institute*

Corresponding author: soorya.elumalai@gmail.com

11.1. Introduction

The uncertainty of global climate change and emerging trends in population growth were the major threat to global food security. Climate change adversely affects both quality and quantity of wheat and rice crops. Crop models predicted yield reduction of 17.2% and 12% in rice and wheat due to the antagonistic effects of climate change [1]. Yield equation comprises of two components namely Harvest index and biomass. Genetic potential of harvest index in many rice and wheat genotypes has reached plateau of approx. 0.6. Prevailing breeding approaches focus on improvement in biomass through altering light perception and radiation use efficiency without sacrificing harvest index [2].

11.2. The Green chemistry

Plant photochemistry entails multiple processes in the conversion of inorganic CO_2 to organic energy-rich carbohydrates via the use of photons and water. Excitation of chlorophylls in the light harvesting complex and reaction center is the first phase, followed by oxidation of water by the Oxygen evolving complex. The excited chlorophylls employed to transfer electrons through photosystem II, Cytochrome b6f, and photosystem I in thylakoids result in the generation of ATP and NADPH. These light-produced compounds are employed in carbon reactions to fix CO_2 in the stromal matrix with Rubisco [3]. Improving photosynthesis focused on several elements, most notably Rubisco bioengineering, CO_2 enrichment near Rubisco, effective perception and utilization of light, and efficient water use. Based on the preceding viewpoint, this chapter discusses numerous ways used to boost photosynthesis and carbon uptake in plants.

11.3. Enhancement of CO_2 Inside the Leaf

11.3.1. Stomatal Conductance (GS)

CO_2 diffusion encounters many barriers along the road, one of which is stomatal conductance, which is a substantial impediment to boosting photosynthetic efficiency. It allows gases such as CO_2 and water vapour to exchange between the leaf and the atmosphere. Stomatal conductance generally depicts as $gs = A/c_a - c_i$. Increasing gs directly improves assimilation rate. gs is affected by a variety of environmental and plant variables [4]. An inevitable environmental component impacting gs is vapour pressure deficit whereas controllable variables are stomatal density, guard cell size, and stomatal response speed. Variation in SD has been well established in genetic improvement of gs. Stomatal density can be genetically controlled by increasing or decreasing the expression of certain transcription factors such as epidermal patterning factor, stomagen, and stomatal density distribution (SDD1) etc [5]. Stomatal kinetics have an eminent role in boosting gs through rapid stomatal opening caused by small stomatal size, and increased density which considerably improves photosynthetic efficiency [6]. Guard cell kinetics is regulated by membrane voltage which is a coordinated function of ion transporters in both plasma membrane and tonoplast. Manipulation of guard cell turgor either through optogenetics or engineering of native stomatal membrane transport such as H^+-ATPase causes enhanced stomatal opening and promoted carbon assimilation [7].

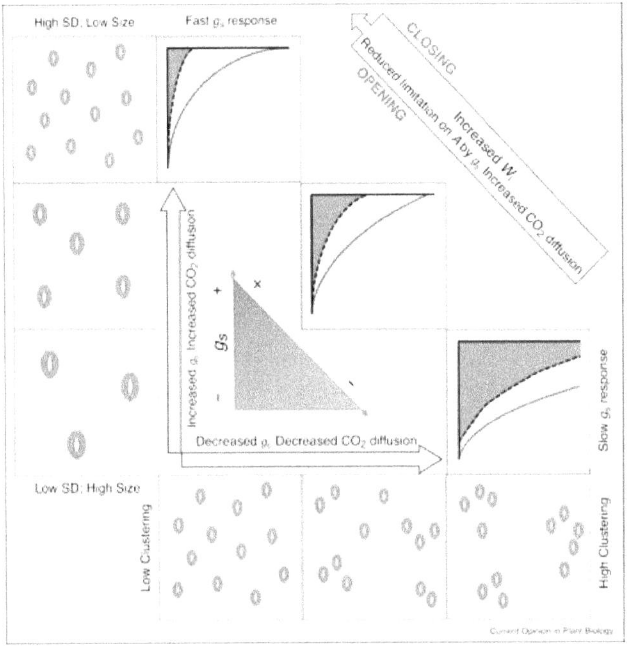

Variation in stomatal density, size and response in relation to stomatal conductance and CO_2 diffusion

11.3.2. Mesophyll conductance (gm)

CO_2 diffusion inside the leaf is restrained by liquid phase resistance. Unlike stomatal conductance which simply restricts diffusion through stomatal pores, mesophyll conductance is a multifactorial parameter that combines cell wall, plasma membrane, cytoplasm and chloroplast envelope [8]. Surface area of the chloroplast facing intercellular space (S_c) determines the efficiency of CO_2 diffusion in mesophyll conductance. Xiong et al., [9] stated that mutants with bigger chloroplasts had considerably lower Sc, which had a negative effect on gm when compared to wild type Arabidopsis plants with smaller chloroplasts. Cell walls are complex structures filled with apoplastic fluid and contain pores which paves a way for liquid phase CO_2 diffusion. Cell wall resistance is determined by several elements, including cell wall thickness, porosity, tortuosity, and CO_2 diffusivity in apoplastic fluid. The thickness and composition of the cell wall have a significant impact on mesophyll conductivity, whereas other factors are subject to estimation uncertainty [10]. A double membrane surrounds both the plasma membrane and the chloroplast. These lipid bilayers are made up of various integral proteins, which may inhibit bulk CO_2 diffusion, but certain integral proteins, such as aquaporins, promote facilitated CO_2 transport. RNAi studies in siencing NtAQP1, a aquaporin localized in both chloroplast inner envelope and plasma membrane showed 90% CO_2 permeability reduction by chloroplast envelope compared to only 10% reduction in plasma membrane [11]. Carbonic Anhydrase activity is found in the stroma of C3 plants, which aids in fascillitated CO_2 diffusion, and it has been hypothesised that removing stromal CAs will dramatically limit mesophyll conductance [12]. Mesophyll conductance is one of the key traits to be focused on improving photosynthesis by increased CO_2 diffusion.

11.4. Enhancement of CO_2 near Rubisco viscinity

11.4.1. Integration of Carboxysomes

The enzyme Rubisco catalyses the interaction of ribulose-1,5-bisphosphate (RuBP) with ambient CO_2, resulting in carbon fixation into organic compounds. It can also catalyse a reaction with oxygen, resulting in photorespiration. It results in the loss of fixed CO_2 and a reduction in photosynthetic efficiency [13]. In cyanobacteria, Rubisco is less likely to react with oxygen when it is encapsulated in a protein microcompartment known as a carboxysome and surrounded by larger quantities of CO_2 [14]. This is a great approach for improving photosynthesis. A transgenic construct must have a 5′ untranslated region (UTR) that provides translation signals and a terminator region for RNA transcript stabilisation in order to produce a foreign gene from a chloroplast genome [15]. A single promoter can drive the transcription of two or more chloroplast genes in a polycistronic transcript. An Intercistronic Expression Element (IEE) with a nuclease-recognized sequence can be put between distinct transgenes on a polycistronic transcript to increase the chance of efficient translation [16]. Typically, chloroplast transformants (transplastomic plants) are created using particle bombardment with a biolistic apparatus. Because of their great regenerability and quick growth in culture, Nicotiana species continue to be the best model system for chloroplast transformation [17].

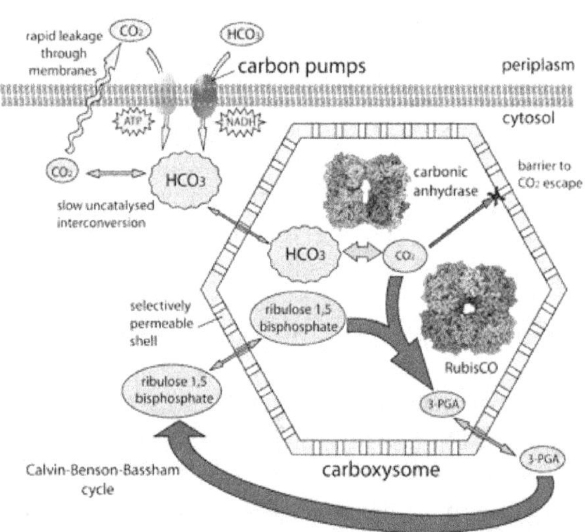

11.4.2. Integration C4 CCM in C3 Plants

In general C4 plants are photosynthetically efficient than C3 plants in rising global temperature due to the spatial difference of carbon fixation and calvin cycle. Bioengineering of C3 plants in to C4 plants is the undergoing strategy to improve photosynthetic efficiency. Phosphoenol pyruvate carboxylase (PEPC), pyruvate orthophosphate dikinase (PPDK), and carbonic anhydrase (CA) are expressed only in mesophyll cells in C4 plants, whereas RuBisCO, and NADP-ME are expressed exclusively in the BS. Most of these genes are also expressed in C3 plants, but in a non-specific manner. To integrate C4 in C3, the expression pattern of these genes must be altered which require Mesophyll- and bundle-sheath-cell-specific promoters [19]. C4 leaf is frequently made up of a repeating pattern of vein-BS-M-M-BS-vein. BSCs encircled by MCs according to kranz anatomy. Hence close interaction between MC and BSC is required for an effective C4 pathway [20]. Given that C3 plants BSCs have fewer chloroplasts with smaller sizes, they do not contribute more to photosynthesis. Increasing the quantity and size of chloroplasts in bundle sheath cells is a successful technique for converting C3 to C4 plants [21]. Different molecular biology tools such as comparative genomics, Crispr Cas 9 and gene stacking will greatly reduce number of transgenes and allow coordinated gene expression.

11.4.3. Utilizing Photorespiratory Bypass to Efficiently Use CO_2

We already covered Rubisco oxygenation, which results in a mechanism called as photorespiration. In photorespiration, 2-PG is a hazardous compound that inhibits chloroplast photosynthesis. To avoid this negative consequence, 2-PG is converted to glycolate, which is delivered to the peroxisomes, where the two molecules of 2-PG finally combine to generate one molecule of 3-PGA with the loss of one molecule of CO_2 via mitochondria [22]. Previously, it was assumed that eliminating photorespiration would significantly improve carbon fixation. Later photorespiratory mutants demonstrated that mutations of genes encoding proteins in the photorespiratory core cycle and related processes are fatal in normal air but viable in increased CO_2 environments [23]. Photorespiratory bypass is one strategy which effectively reduces the harmful effect of photorespiration by releasing CO_2 directly in to the chloroplast instead of mitochondria.Major photorespiratory bypass in current focus constitutes the bacterial glycolate dehydrogenase (GDH) which oxidise glycolate to glyoxylate in the chloroplasts. Then, glyoxylate carboxyligase (GCL) converts two molecules of the 2-carbon glyoxylate to a three-carbon tartronic semialdehyde (TSA), releasing one CO_2 inside the chloroplast. To re-enter photosynthesis, tartronic semialdehyde is converted to glycerate by the enzyme tartronic semialdehyde reductase (TSR) [24]. Transgenic breeding to establish photorespiratory bypass in C3 plants greatly enhance carbon fixation.,Red arrows indicate photorespiratory bypass and enzymes involved in it.

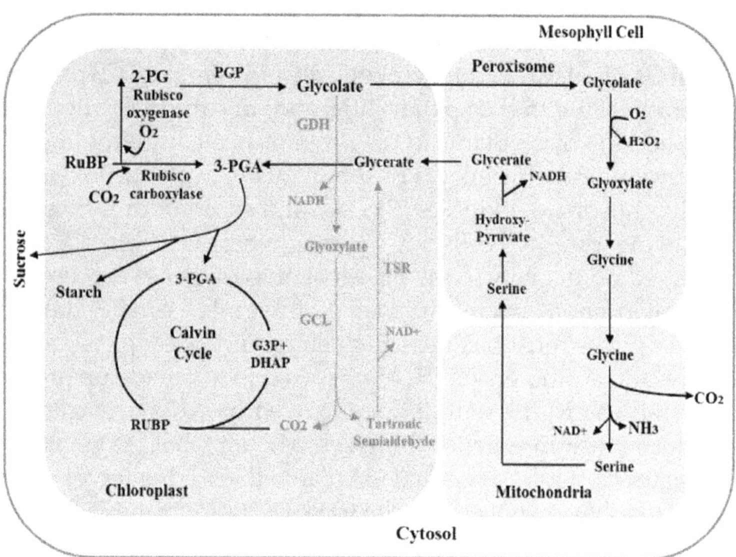

11.4.4. Bioengineering of Rubisco

Rubisco is the most important nitrogen consumer in the plant system. Because nitrogen is a key constraint and a costly nutrient, reducing nitrogen overinvestment in rubisco is obviously a primary priority for improving nutrient usage efficiency. Increasing the quantity of rubisco may boost photosynthetic efficiency to some level, however studies have shown that the favourable effect of greater nitrogen investment in rubisco is offset by the photorespiratory process [25]. As a result, the goal of photosynthetic breeding is to promote rubisco activity and selectivity rather than rubisco abundance. Natural diversity in the L8S8 components of Rubisco reveals that Rubisco evolved differently under diverse environmental conditions. Some Rubisco isomers, such as red L8S8, developed quicker in crop plants than green L8S8, and these rubisco have better CO_2/O_2 selectivity than plant rubisco. The incompatibility of molecular chaperones, which is necessary in rubisco synthesis, is a problem in transforming these red rubisco from red algae [26]. Though Rubisco's active sites are found on the large subunits, expression of the small subunit governs the amount of the Rubisco pool in plants which can influence the overall catalytic effectiveness of the Rubisco complex. The small subunit is one of the possible engineering target for improving Rubisco performance [27]. Rubisco activase is a crucial enzyme that carboxylates and activates rubisco in an ATP-dependent manner when exposed to light. Rubisco activase is more heat sensitive than Rubisco, hence creating thermotolerance rubisco activase will significantly enhance photosynthetic efficiency under rising temperatures [28]. Rubisco from archae contains RbCL 2 subunit, which has a strong affinity for oxygen. This information aids in the identification of a specific amino acid residue, methionine 295 in RbCL 2, which is critical for the oxygenation of form 1 rubisco in higher plants.

Recombinant Met-295 was less sensitive to oxygen and had a higher CO_2/O_2 specificity. Thus, manipulating particular amino acids in rubisco is one way for improving carbon capture efficiency [29].

11.4.5. Engineering of Calvin Cycle

Both the ability of RuBP to regenerate and the ability of Rubisco to carboxylate are necessary for the photosynthetic fixation of CO_2. Three unregulated enzymes, fructose 1,6-bisphosphate aldolase (aldolase), sedoheptulose 1,7-bisphosphatase (SBPase), and transketolase (TK), have significantly influence RuBP regeneration. This suggests that they have substantial influence over photosynthetic carbon flow and can constrain photosynthetic rate in addition to Rubisco. In order to increase the capacity for photosynthetic production, these three enzymes may represent possible engineering targets [30]. For instance, tobacco plants greatly enhanced their photosynthesis and growth when either cyanobacterial fructose 1,6-/sedoheptulose 1,7-bisphosphatase (FBP/SBPase) or plant SBPase were overexpressed [31].

11.5. Manipulation of Electron Transport Chain

Similar to RUBP regeneration, a decrease in the electron transport chain reduces photosynthesis since it generates ATP and NADPH, both of which are necessary in CO_2 reduction. According to research, boosting cytochrome b6f activity may significantly boost photosynthetic rate. The multiprotein subunits and complicated structural makeup of cytb6f make manipulation difficult. When combined with the rieske Fe-S complex, cytochrome B6 performs as a dimer. Overexpression of Rieske FeS enhances the abundance of functional cytochrome b6f and may have the potential to increase plant productivity if combined with other traits [32]. The light harvesting complex is an auxiliary component that surrounds PSI and PS II, absorbing light and transferring energy to the reaction center. Findings suggests that reducing the size of the light-harvesting antennas in tobacco photosystems helps to boost photosynthetic output and plant canopy biomass accumulation under high-density farming circumstances. The Truncated Light Harvesting chlorophyll Antenna size idea has already been established in green microalgae and cyanobacteria, but it has yet to be demonstrated in agricultural plants [33]. High light conditions frequently damage photosensitive thylakoids; this additional energy is released as heat via the xanthophyll cycle, a process known as non-photochemical quenching. The surplus light energy is dissipated as heat during the conversion of violaxanthin to zeaxanthin by violaxanthin de-epoxidase. This photoprotection technique lasts for many minutes, even in low light situations. In changing light, an accelerated response to natural shade boosted leaf carbon dioxide absorption and plant dry matter output by nearly 15% [34]. Thus focus on rapid shut down off NPQ system by over expression of Zeaxanthin epoxidase under shade condition is the trait of intrest for crop improvement breeding programs.

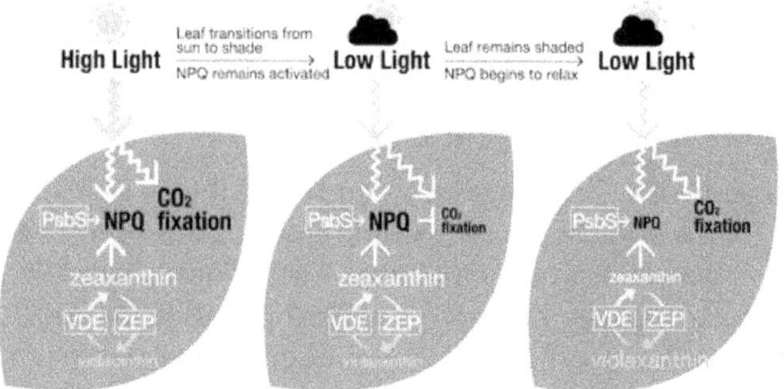

Representation of rapid Non photochemical quenching (NPQ) under high light and slow relaxation of NPQ under subsequent lowlight conditions

11.6. Improving Radiation use Efficiency and Canopy Photosynthesis

The quantity of carbon fixation per unit of photosynthetically active radiation (PAR) absorbed on crop canopy is referred to as radiation utilization efficiency. RUE is improved by increasing leaf area, total chlorophyll, source-sink relationships, and photosynthetic metabolism. Light varies throughout the crop canopy under natural situations. The top leaves receive more light, while the bottom leaves are shaded. According to the findings, lower leaves have a higher absorption capacity and a lower light reflectance [35]. Breeding for canopy architecture by engineering leaf angles is the current focus in improving rice productivity. Canopy photosynthesis is not similar to leaf photosynthesis due to the varied photon flux density across the canopy heights. Comparative analysis of rice genotypes indicates that differing in chlorophyll content does not correlate with CO_2 fixation capacity [36]. Accurate quantification of canopy photosynthesis is difficult due to the changing light intensities, however canopy photosynthetic models such as artificial neural network had been developed to estimate canopy photosynthesis using single leaf photosynthesis and leaf area index (LAI) [37]. Developing a canopy in which upper leaf is more vertical with shorter antenna outside the thylakoids compared to lower leaves will greatly improves canopy photosynthesis and radiation use efficiency [38].

11.7. Conclusion

Improving photosynthetic efficiency is a complicated dynamic collection of operations that incorporates many pathways ranging from light interception in thylakoids to CO_2 fixation in stroma. The efficiency of photosynthesis is determined by a variety of external parameters such as vapour pressure deficit, canopy temperature, photosynthetic photon flux density, CO_2 availability, and so on. Carbon fixation may be improved by changing parameters such as Rubisco specificity, stomatal conductance, mesophyll conductance, canopy design, inclusion of cyanobacterial and algal rubiscos and CCMs, and so on. These tactics are realised through the development of molecular breeding tools such as Crispr Cas, mutational breeding, transformation research, and gene stacking, among others. In addition to these molecular breeding procedures, numerous new crop models aid in improving crop plant photosynthetic efficiency under changing global climatic circumstances.

Reference

Habib-ur-Rahman, M., Ahmad, A., Raza, A., Hasnain, M. U., Alharby, H. F., Alzahrani, Y. M., ... & El Sabagh, A. (2022). Impact of climate change on agricultural production; Issues, challenges, and opportunities in Asia. *Frontiers in Plant Science, 13*, 925548.

Furbank, R. T., Sharwood, R., Estavillo, G. M., Silva-Perez, V., & Condon, A. G. (2020). Photons to food: genetic improvement of cereal crop photosynthesis. *Journal of Experimental Botany, 71*(7), 2226-2238.

van Bezouw, R. F., Keurentjes, J. J., Harbinson, J., & Aarts, M. G. (2019). Converging phenomics and genomics to study natural variation in plant photosynthetic efficiency. *The Plant Journal, 97*(1), 112-133.

Wang, Y., Wang, Y., Tang, Y., & Zhu, X. G. (2022). Stomata conductance as a goalkeeper for increased photosynthetic efficiency. *Current Opinion in Plant Biology*, 102310.

Faralli, M., Matthews, J., & Lawson, T. (2019). Exploiting natural variation and genetic manipulation of stomatal conductance for crop improvement. *Current Opinion in Plant Biology, 49*, 1-7.

Drake, P. L., Froend, R. H., & Franks, P. J. (2013). Smaller, faster stomata: scaling of stomatal size, rate of response, and stomatal conductance. Journal of experimental botany, 64(2), 495-505.

Nguyen, T. B. A., Lefoulon, C., Nguyen, T. H., Blatt, M. R., & Carroll, W. (2023). Engineering stomata for enhanced carbon capture and water-use efficiency. Trends in Plant Science.

Evans, J. R. (2021). Mesophyll conductance: walls, membranes and spatial complexity. New Phytologist, 229(4), 1864-1876.

Xiong, D., Huang, J., Peng, S., & Li, Y. (2017). A few enlarged chloroplasts are less efficient in photosynthesis than a large population of small chloroplasts in Arabidopsis thaliana. Scientific reports, 7(1), 5782.

Roig-Oliver, M., Bresta, P., Nadal, M., Liakopoulos, G., Nikolopoulos, D., Karabourniotis, G., ... & Flexas, J. (2020). Cell wall composition and thickness affect mesophyll conductance to CO_2 diffusion in Helianthus annuus under water deprivation. Journal of Experimental Botany, 71(22), 7198-7209.

Uehlein, Norbert, Beate Otto, David T. Hanson, Matthias Fischer, Nate McDowell, and Ralf Kaldenhoff. "Function of Nicotiana tabacum aquaporins as chloroplast gas pores challenges the concept of membrane CO2 permeability." The Plant Cell 20, no. 3 (2008): 648-657.

Tholen, D. and Zhu, X.G., 2011. The mechanistic basis of internal conductance: a theoretical analysis of mesophyll cell photosynthesis and CO2 diffusion. Plant physiology, 156(1), pp.90-105.

Parry, M.A., Andralojc, P.J., Scales, J.C., Salvucci, M.E., Carmo-Silva, A.E., Alonso, H. and Whitney, S.M., 2013. Rubisco activity and regulation as targets for crop improvement. Journal of experimental botany, 64(3), pp.717-730.

Badger, M. R., Hanson, D., & Price, G. D. (2002). Evolution and diversity of CO2 concentrating mechanisms in cyanobacteria. Functional Plant Biology, 29(3), 161-173.

Yang, H., Gray, B. N., Ahner, B. A., & Hanson, M. R. (2013). Bacteriophage 5ᵁ untranslated regions for control of plastid transgene expression. Planta, 237, 517-527.

Zhou, F., Karcher, D., & Bock, R. (2007). Identification of a plastid intercistronic expression element (IEE) facilitating the expression of stable translatable monocistronic mRNAs from operons. The Plant Journal, 52(5), 961-972.

Maliga, P., & Tungsuchat-Huang, T. (2014). Plastid transformation in Nicotiana tabacum and Nicotiana sylvestris by biolistic DNA delivery to leaves. Chloroplast biotechnology: methods and protocols, 147-163.

Espie, G. S., & Kimber, M. S. (2011). Carboxysomes: cyanobacterial RubisCO comes in small packages. Photosynthesis Research, 109, 7-20.

Weber, A. P., & von Caemmerer, S. (2010). Plastid transport and metabolism of C3 and C4 plants—comparative analysis and possible biotechnological exploitation. Current opinion in plant biology, 13(3), 256-264.

Feldman, A. B., Leung, H., Baraoidan, M., Elmido-Mabilangan, A., Canicosa, I., Quick, W. P., ... & Murchie, E. H. (2017). Increasing leaf vein density via mutagenesis in rice results in an enhanced rate of photosynthesis, smaller cell sizes and can reduce interveinal mesophyll cell number. Frontiers in Plant Science, 8, 1883.

Gowik, U., & Westhoff, P. (2011). The path from C3 to C4 photosynthesis. Plant Physiology, 155(1), 56-63.

Peterhansel, C., Horst, I., Niessen, M., Blume, C., Kebeish, R., Kürkcüoglu, S., & Kreuzaler, F. (2010). Photorespiration. The Arabidopsis Book/American Society of Plant Biologists, 8.

Timm, S., & Bauwe, H. (2013). The variety of photorespiratory phenotypes–employing the current status for future research directions on photorespiration. Plant Biology, 15(4), 737-747.

Dalal, J., Lopez, H., Vasani, N. B., Hu, Z., Swift, J. E., Yalamanchili, R., ... & Sederoff, H. W. (2015). A photorespiratory bypass increases plant growth and seed yield in biofuel crop Camelina sativa. Biotechnology for biofuels, 8, 1-22.

Zhu, X. G., De Sturler, E., & Long, S. P. (2007). Optimizing the distribution of resources between enzymes of carbon metabolism can dramatically increase photosynthetic rate: a numerical simulation using an evolutionary algorithm. Plant physiology, 145(2), 513-526.

Oh, Z. G., Askey, B., & Gunn, L. H. (2023). Red Rubiscos and opportunities for engineering green plants. Journal of Experimental Botany, 74(2), 520-542.

Mao, Y., Catherall, E., Díaz-Ramos, A., Greiff, G. R., Azinas, S., Gunn, L., & McCormick, A. J. (2023). The small subunit of Rubisco and its potential as an engineering target. Journal of Experimental Botany, 74(2), 543-561.

Scafaro, A. P., Atwell, B. J., Muylaert, S., Reusel, B. V., Ruiz, G. A., Rie, J. V., & Gallé, A. (2018). A thermotolerant variant of Rubisco activase from a wild relative improves growth and seed yield in rice under heat stress. Frontiers in Plant Science, 9, 1663.

Kreel, N. E., & Tabita, F. R. (2007). Substitutions at methionine 295 of Archaeoglobus fulgidus ribulose-1, 5-bisphosphate carboxylase/oxygenase affect oxygen binding and CO2/O2 specificity. Journal of Biological Chemistry, 282(2), 1341-1351.

Raines, C. A. (2003). The Calvin cycle revisited. Photosynthesis research, 75, 1-10.

Tamoi, M., Nagaoka, M., Miyagawa, Y., & Shigeoka, S. (2006). Contribution of fructose-1, 6-bisphosphatase and sedoheptulose-1, 7-bisphosphatase to the photosynthetic rate and carbon flow in the Calvin cycle in transgenic plants. Plant and cell physiology, 47(3), 380-390.

Simkin, A. J., McAusland, L., Lawson, T., & Raines, C. A. (2017). Overexpression of the RieskeFeS protein increases electron transport rates and biomass yield. Plant physiology, 175(1), 134-145.

Kirst, H., Gabilly, S. T., Niyogi, K. K., Lemaux, P. G., & Melis, A. (2017). Photosynthetic antenna engineering to improve crop yields. Planta, 245, 1009-1020.

Kromdijk, J., Głowacka, K., Leonelli, L., Gabilly, S. T., Iwai, M., Niyogi, K. K., & Long, S. P. (2016). Improving photosynthesis and crop productivity by accelerating recovery from photoprotection. Science, 354(6314), 857-861.

Wang, B., Smith, S. M., & Li, J. (2018). Genetic regulation of shoot architecture. Annual review of plant biology, 69, 437-468.

WEI, H. H., YANG, Y. L., SHAO, X. Y., SHI, T. Y., MENG, T. Y., Yu, L. U., ... & DAI, Q. G. (2020). Higher leaf area through leaf width and lower leaf angle were the primary morphological traits for yield advantage of japonica/indica hybrids. Journal of Integrative Agriculture, 19(2), 483-494

Kaneko, T., Nomura, K., Yasutake, D., Iwao, T., Okayasu, T., Ozaki, Y., ... & Kitano, M. (2022). A canopy photosynthesis model based on a highly generalizable artificial neural network incorporated with a mechanistic understanding of single-leaf photosynthesis. Agricultural and Forest Meteorology, 323, 109036.

Ort, D. R., Merchant, S. S., Alric, J., Barkan, A., Blankenship, R. E., Bock, R., ... & Zhu, X. G. (2015). Redesigning photosynthesis to sustainably meet global food and bioenergy demand. Proceedings of the national academy of sciences, 112(28), 8529-8536.

Chapter–12

Innovative Plant Genomic Approaches in Breeding Disease Resistant Crop

Poojashree Mahapatra[1] and Tushar Arun Mohanty[2*]

1. *Department of Plant Pathology, College of Agriculture, Odisha University of Agriculture & Technology, Odisha.*

2. *Department of Genetics and Plant Breeding), Centre for Plant Breeding and Genetics, TNAU, Coimbatore.*

**Corresponding author: tushararunmohanty@gmail.com*

12.1 Introduction

Modern agriculture faces a critical challenge in meeting the food needs of a rapidly growing global population, projected to reach over 9.8 billion by 2050 from 7.3 billion in 2015. This challenge is compounded by significant crop losses caused by diseases. Bacterial and fungal pathogens alone reduce crop yields by approximately 15%, while viruses contribute to a 3% yield decrease. Pathogens, including fungi, oomycetes, bacteria, and viruses, are major obstacles to agricultural productivity and quality (Li *et al.* 2020)

To address this challenge, the development of crop varieties with strong, long-lasting, and broad-spectrum disease resistance is an economically and environmentally sustainable alternative to costly and environmentally harmful chemical control methods (Mahlein *et al.* 2019 and Piquerez *et al.* 2014). Niks *et al.* have categorized host resistance based on its impact on the phenotype and mode of inheritance. Qualitative resistance, associated with single-gene inheritance and often following the gene-for-gene interaction model, provides nearly complete defense but only against specific pathogen genotypes. While useful in plant breeding, qualitative resistance is limited in duration as pathogens adapt to overcome it. In contrast, quantitative genetic resistance, controlled by multiple genes or Quantitative Trait Loci (QTL), displays complex multigenic inheritance, posing challenges for breeding efforts (Pilet-Nayel *et al.* 2017).

Traditional breeding programs have made substantial progress over the past century but have been constrained by long breeding cycles and the time required from initial crosses to the release of new cultivars (Hickey *et al.* 2017). Genes linked to pathogen tolerance or resistance can be sourced from local germplasm, exotic lines, wild species, or other breeding programs (Varshney *et al.* 2011)

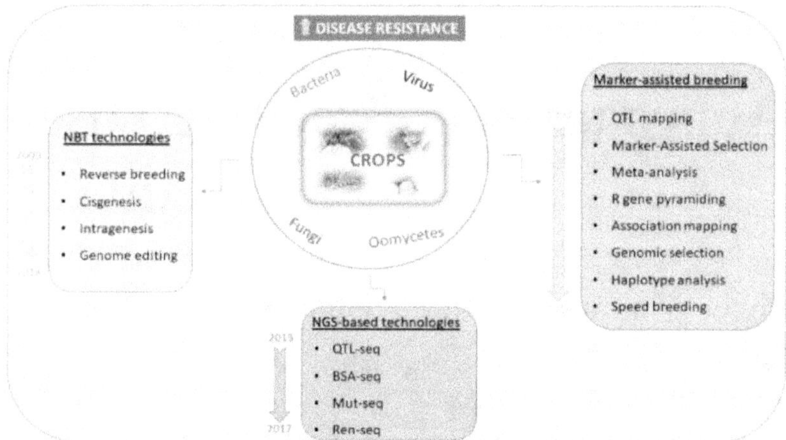

Figure 1. Schematic representation of genetics and "omics" approaches to study and solve disease resistance in crops, reviewed in this report (NBT: New Breeding Technologies; QTL: Quantitative Trait Locus; BSA: Bulk Segregant Analysis; Mut: Mutant; Ren: R gene enrichment; R: Resistance) (Mores *et al.* 2021).

The introduction of molecular markers has transformed plant breeding by providing a practical means of selecting genes. In the post-genomic era, access to genomic tools and genetic resources has given rise to innovative methods in plant breeding, particularly for complex traits. Genomic approaches have revolutionized the development of disease-resistant crops, offering a path to sustainable agriculture. By identifying genes and genetic markers linked to resistance, these techniques enable breeders to precisely and swiftly develop crop varieties with enhanced disease resilience. This precision expedites breeding processes and enables targeted protection against specific pathogens, enhancing the effectiveness of disease prevention strategies. Additionally, genomic tools reduce the environmental impact of agriculture by decreasing reliance on chemical pesticides and promoting eco-friendly farming practices. These tools empower farmers to address evolving disease challenges and optimize crop management, contributing to long-term agricultural resilience and global food security in a changing world.

This chapter provides a comprehensive overview of molecular approaches for identifying molecular markers, both on a genome-wide scale and closely linked to resistance genes, for genotypic characterization in crop breeding. It covers a range of strategies, from Marker-Assisted Selection (MAS) and Genomic Selection to Machine Learning and genetic engineering. The practical applications of these approaches are explored in the context of their potential integration into breeding programs to enhance plant disease resistance in crops.

12.2. Understanding the genetic basis of disease resistance

Understanding the genetic basis of disease resistance in plants is pivotal in agriculture and plant science. It involves the identification of specific genes or genetic markers responsible for conferring resistance to diseases caused by pathogens like fungi, bacteria, viruses, and nematodes. This knowledge empowers breeders to select and manipulate these genes to develop disease-resistant crop varieties. Moreover, genetic research quantifies resistance levels, enabling breeders to assess a variety's protective capabilities and make informed breeding decisions. With this understanding, breeders can employ advanced techniques like marker-assisted selection and genome editing, expediting the development of disease-resistant crops. Additionally, insight into the mechanisms underlying plant-pathogen interactions can lead to innovative strategies for enhancing resistance, such as genetic engineering and biocontrol agents. Furthermore, as pathogens evolve and new diseases emerge, this genetic knowledge becomes crucial for adapting crops rapidly. Disease-resistant crops also often require fewer chemical pesticides, reducing agriculture's environmental impact and contributing to sustainable farming practices. Ultimately, unraveling the genetic basis of disease resistance is indispensable for food security and sustainable agriculture in an ever-changing world. The genetic basis of disease resistance in plants is a complex and multifaceted process that involves various genetic components and mechanisms. Here is an overview of the key elements that contribute to disease resistance in plants:

12.2.1 R Genes (Resistance Genes)

R genes are specific genes within a plant's genome that encode for proteins involved in recognizing and responding to pathogens. These proteins are receptors that can detect pathogen-derived molecules, known as avirulence (Avr) proteins or effectors, which are released by the pathogen during infection. When an R gene recognizes a compatible effector, it triggers a defense response in the plant, often leading to resistance against the pathogen.

12.2.2. Effector Recognition

Pathogens secrete effectors to manipulate the plant's immune system and facilitate infection. Plants have evolved R genes that can recognize these effectors, initiating a cascade of molecular events that activate defense mechanisms.

12.2.3 Signal Transduction Pathways

Once an R gene is triggered, it activates a series of intracellular signal transduction pathways. These pathways involve the activation of various enzymes and the production of signaling molecules like salicylic acid, jasmonic acid, and ethylene, which coordinate the plant's defense responses.

12.2.4. Cell Wall Reinforcement

One common defense mechanism involves strengthening the plant's cell wall through the deposition of lignin and other structural molecules. This physical barrier makes it harder for pathogens to penetrate plant cells.

12.2.5. Production of Antimicrobial Compounds

Plants can synthesize antimicrobial compounds, such as phytoalexins and pathogenesis-related (PR) proteins, in response to pathogen attack. These compounds inhibit the growth of pathogens or interfere with their life cycles.

12.2.6. Programmed Cell Death (Hypersensitive Response)

In some cases, plants activate a form of programmed cell death called the hypersensitive response (HR) at the site of pathogen invasion. This localized cell death prevents the pathogen from spreading to other parts of the plant.

12.2.7. Systemic Acquired Resistance (SAR)

SAR is a defense mechanism where the plant, following an initial pathogen encounter, activates a long-lasting systemic response throughout its tissues. This systemic resistance makes the entire plant more resistant to subsequent infections.

12.2.8. Quantitative Resistance

In addition to the major R genes, many plants also exhibit quantitative resistance, which involves multiple genes with smaller, additive effects on disease resistance. This polygenic trait contributes to a broader and more durable resistance against pathogens.

12.2.9. Epigenetic Regulation

Epigenetic modifications, such as DNA methylation and histone modifications can play a role in regulating gene expression related to disease resistance.

Plant immune responses and resistance genes play a pivotal role in defending plants against pathogens and pests, contributing to their survival and productivity. Here's an overview of their roles and how they function:

12.3. Plant Immune Responses

Plant immune responses are the defensive mechanisms that plants employ to recognize and combat invading pathogens, including fungi, bacteria, viruses, and herbivores. These responses can be broadly categorized into two major branches: the basal or non-host resistance and the specific or gene-for-gene resistance.

12.3.1. Basal (Non-host) Resistance

Basal resistance is the plant's general defense mechanism against a wide range of pathogens that it does not typically host. It involves physical barriers like the plant's cell wall and the production of antimicrobial compounds such as phytoalexins. Basal resistance is always active, providing a basic level of protection.

12.3.2 Gene-for-Gene (Specific) Resistance

In cases where a plant and a pathogen interact more closely, such as in compatible host-pathogen interactions, specific resistance mechanisms come into play. These mechanisms rely on the recognition of pathogen molecules (effectors) by plant receptors (R genes). When an R gene recognizes a specific effector, it triggers a hypersensitive response (HR), which includes programmed cell death at the site of infection, preventing the pathogen from spreading further.

12.4. Resistance Genes (R Genes)

Resistance genes, also known as R genes, are a critical component of the plant's immune system. They are plant genes that encode receptors capable of recognizing pathogen effectors or other molecules associated with pathogens. Here's how R genes function:

12.4.1. Effector Recognition

R genes encode receptor proteins that can recognize specific pathogen effectors. These effectors are proteins produced by pathogens to suppress plant defenses or promote infection.

12.4.2. Triggering Defense Responses

When an R gene recognizes the presence of a compatible effector, it initiates a signaling cascade within the plant cell. This cascade leads to the activation of defense mechanisms, such as the production of antimicrobial compounds, reinforcement of the cell wall, and systemic acquired resistance (SAR).

12.4.3. Variability and Specificity

R genes are highly variable among plant species and can be specific to certain pathogens or pathogen strains. This specificity is a key aspect of gene-for-gene resistance.

12.4.4. Breeding for Resistance

In agriculture, plant breeders often use R genes to develop disease-resistant crop varieties through traditional breeding or, more recently, through genetic engineering techniques. These resistant varieties can provide reliable protection against specific pathogens, reducing the need for chemical pesticides.

12.5. Traditional breeding methods

Traditional breeding methods have been used for centuries to develop disease-resistant crops by selecting and crossing plants with desirable traits. Here's an overview of the steps and techniques involved in traditional breeding for disease resistance:

12.5.1. Selection of Parental Plants

Breeders start by identifying plants within a crop species that exhibit some level of resistance to the target disease. These plants serve as the parental or donor plants for the breeding program. Ideally, they possess genetic traits that confer resistance to the disease of interest.

12.5.2. Crossing (Hybridization)

Breeders then cross the selected parental plants, typically one with resistance and the other with desirable agronomic traits (e.g., high yield, quality, or adaptation to specific growing conditions). This results in the creation of a population of hybrid plants that inherit genetic material from both parents.

12.5.3. Generation of Segregating Populations

The hybrid plants represent the first generation (F1). When these F1 plants self-pollinate or are crossed back with one of the parent lines, a population of F2 plants is produced. This population is genetically diverse, as it contains various combinations of genes from both parental lines.

12.5.4. Phenotypic Selection

The F2 generation and subsequent generations (F3, F4, etc.) are evaluated for disease resistance. Breeders assess the phenotype (observable characteristics) of the plants, including their resistance to the target disease. Plants displaying the desired resistance traits are selected for further breeding.

12.5.5. Backcrossing

In cases where the resistance donor lacks certain desirable agronomic traits or has undesirable characteristics, breeders can perform backcrossing. This involves crossing a selected resistant plant from the F2 or later generations back to the recurrent parent with the desirable agronomic traits. The objective is to transfer the resistance trait while retaining the desirable characteristics of the recurrent parent.

12.5.6. Repetitive Selection

The breeding process continues through multiple generations, with repeated cycles of selection, hybridization, and backcrossing. This helps refine the population, concentrating the genes responsible for disease resistance and other desirable traits.

12.5.7. Field Trials and Evaluation

The selected plant populations are subjected to rigorous field trials and disease assessments in various environments to confirm their resistance and overall performance. Only the best-performing lines are advanced to subsequent generations.

12.5.8. Varietal Release

Once a stable, disease-resistant variety with the desired agronomic traits is achieved, it undergoes official testing and evaluation for varietal release. If it meets the criteria for commercial cultivation, the new crop variety is registered and made available to farmers.

Traditional breeding methods for disease resistance are time-consuming and can take several years to develop a new resistant crop variety. However, they have been highly successful in creating numerous disease-resistant crop varieties across a wide range of crops, contributing significantly to global food security and sustainable agriculture

12.6. Innovative Genomic Approaches

Innovative genomic approaches, such as single-cell genomics and CRISPR-Cas9 technology, are transforming our understanding of genetics and enabling precise DNA editing. Epigenomics techniques reveal intricate gene regulation mechanisms, while metagenomics explores diverse microbial ecosystems. Long-read sequencing enhances genomic analysis, and functional genomics tools like CRISPR screens identify essential genes. Precision medicine tailors treatments to individuals, liquid biopsies aid in cancer diagnosis, and synthetic biology creates novel biological systems. AI and machine learning analyze vast genomic data, while ethical and privacy concerns accompany increased data accessibility. These approaches collectively drive genetics, healthcare, and biotechnology forward, offering substantial promise for scientific discovery and personalized medicine.

12.6.1. Toward the Identification of Resistance Genes/Loci

Traditional mapping has identified disease resistance genes in many crops, aiding genetic improvement. Advanced genomic tools, driven by available genome data, have accelerated gene identification and precise causal gene location. Our review spans traditional to NGS-based mapping approaches and includes meta-QTL analysis for comparing results across experiments and finding consensus QTLs.

12.6.1.1. Traditional and NGS-Enabled Mapping Approaches

In crop research, two main methods are employed to identify genomic regions associated with resistance traits: linkage analysis and genome-wide association studies (GWAS) (St. Clair *et al.* 2010). Linkage analysis often uncovers quantitative resistance loci (QRLs), which can contain hundreds of genes, making it challenging to pinpoint the true causal gene. Recent advances in Next-Generation Sequencing (NGS) have accelerated the identification of QTLs related to agronomic traits, especially disease resistance. High-density SNP arrays, like those for maize, rice, wheat, and other major crops, have emerged thanks to these advancements. Illumina's Infinium and Affymetrix's Axiom are widely used SNP genotyping platforms (Thomson *et al.* 2014).

Genotyping by sequencing (GBS) offers an alternative approach, particularly for orphan crops, while fixed genotyping chips are preferred for structured datasets in ongoing breeding programs. GBS is better suited for detecting true genetic variants in phenotypes with complex genetic architectures. Despite this, SNP arrays can still reveal genetic associations due to linkage disequilibrium (LD) in most crop genomes. Some crops have specific arrays for breeding or diverse germplasm.

Genome-wide association studies (GWAS) offer higher resolution mapping compared to linkage analysis, making it easier to identify candidate genes. In scenarios with fast LD decay or limited genome size, high-density GWAS with abundant SNPs can reduce the number of candidate genes. Haplotype-based association analysis, focusing on SNP combinations inherited together, can enhance QTL detection and mapping precision, especially when the causal locus is unknown.

Haplotype-based analysis has proven valuable for characterizing diversity, fine mapping known resistance loci, validating loci in different populations, and tracing allelic selection in disease resistance studies. However, only a few methods for haplotype-based GWAS exist in plant genomics. Advancements in high-throughput sequencing are expected to create comprehensive haplotype maps, revolutionizing GWAS by improving genotype imputation and enhancing trait mapping precision across various gene mapping projects.

12.6.1.2. NGS-Enabled Fine-Mapping/Cloning Approaches

Several Next-Generation Sequencing (NGS)-based mapping strategies have emerged in recent years for pinpointing and finely mapping disease resistance loci, as well as developing diagnostic markers for breeding purposes. These methods are especially useful for segregating populations and combine traditional bulk segregant analysis (BSA) with sequencing to detect major resistance loci. They are known by various names like QTL-Seq, Seq-BSA, or Indel-Seq, depending on their focus, such as identifying insertions and deletions.

One example, QTL-Seq, streamlines the identification and fine mapping of quantitative trait loci (QTLs) within smaller genomic regions, expediting the discovery of candidate genes linked to specific traits. Initially applied to blast resistance in rice, QTL-Seq has since been utilized in various species, including pigeonpea, groundnut, pepper, tomato, and watermelon for diverse disease resistance traits. In cases where multiple mapping populations share at least one common parent, it is referred to as multiple QTL-Seq (mQTL-Seq), a strategy that leverages genetic diversity to validate QTLs and narrow down their locations for various agronomic traits.

NGS plays a crucial role in these methods by sequencing groups of plants with differing phenotypes and providing insights into the parent genome structure. However, for crops with large and complex genomes, this approach can be costly and require tailored bioinformatics solutions for genome assembly. To address this, techniques like Bulk Segregant RNA Seq (BSR-Seq) rely on whole transcriptome sequencing of contrasting plant groups and have proven effective for rapidly identifying genes and closely linked markers related to target traits. For instance, it was employed to finely map the Yr15 gene responsible for yellow rust resistance in wheat. Another technique, Specific-Locus Amplified Fragment (SLAF)-Seq, simplifies whole genome resequencing by reducing genome complexity but requires a known reference genome and strong bioinformatics support (Ramirez-Gonzalez *et al.* 2015).

NGS technologies have also accelerated the identification of target genes linked to specific traits using forward genetic approaches. For instance, NGS applied to mutant collections has led to methods like MutMap, MutMap Gap, MutMap+, modified MutMap, and SHOREmap (Abe *et al.* 2012). MutMap, for example, identifies genomic regions associated with agronomic traits through chemical mutagenesis, phenotypic selection, and whole-genome sequencing of groups with contrasting phenotypes.

Furthermore, NGS-based approaches like MutChromSeq, which combines chromosome sorting with sequencing, offer a means to reduce genome complexity and isolate candidate genes. MutChromSeq has successfully cloned genes like the wheat powdery mildew resistance gene Pm2 and the barley leaf rust resistance gene Rph1 (Sánchez-Martín et *al.* 2016 and Dracatos *et al.* 2019). Similarly, TArgeted Chromosome-based Cloning via long-range Assembly (TACCA) combines genome complexity reduction with chromosome sorting and has been used to clone the wheat leaf rust resistance gene Lr22a (Thind *et al.* 2017).

Additionally, NGS-based approaches like RenSeq leverage the abundance of nucleotide binding and leucine-rich repeat (NLR)-encoding genes in plant genomes. RenSeq involves capturing fragments of these genes from genomic libraries and sequencing, offering a cost-effective gene enrichment strategy for identifying and cloning new disease resistance genes. MutRenSeq combines this approach with the comparison of R gene complements between mutants and wild-type progenitors, facilitating the rapid cloning of resistance genes like Sr22 and Sr45 in wheat (Steuernagel *et al.* 2016).

These approaches provide valuable tools for identifying and cloning disease resistance genes in crops. However, further research is necessary to uncover additional resistance genes and loci that can be applied in marker-assisted selection (MAS) and genome editing to enhance crop cultivars.

12.6.1.3 Meta-QTL Analysis for Disease Resistance in Crops

Efficient Marker-Assisted Selection (MAS) requires specific conditions, including the presence of QTLs that explain a substantial portion of observed phenotypic variation and precise genomic localization. A comparative analysis of genomic regions responsible for a trait across various studies, genetic backgrounds, and environments is crucial. This involves projecting QTLs onto a unified genetic map or the genome, identifying consensus QTLs (meta-QTLs) with more precise confidence intervals (CIs), and aligning them with reference genomes to uncover candidate genes.

Numerous studies since 2009 have conducted meta-QTL analyses for resistance against pathogens in various crops. For example, a study in maize identified 104 resistance meta-QTLs from 382 individual QTLs, some effective against multiple insect pests. In wheat, an analysis of Fusarium head blight resistance produced 65 meta-QTLs from 556 individual QTLs. These meta-QTLs were linked to candidate genes and a well-known FHB resistance locus (Fhb1) [91]. Similarly, studies in wheat and common beans found narrower CIs for MQTLs, facilitating marker development for breeding (Cai *et al.* 2019).

Advancements in genome sequencing, like durum wheat, enable projecting QTLs directly onto the genome, enhancing MQTL analysis. In summary, meta-QTLs refine genomic intervals, aiding the search for genes responsible for disease resistance.

12.6.2. Marker-Assisted Selection

Breeding programs rely on assessing genetic diversity within a crop species, identifying advantageous alleles that enhance specific traits, encouraging genetic recombination with elite genotypes, and selecting individuals with favorable phenotypic characteristics. In recent years, molecular marker-assisted selection (MAS) has transformed breeding by shifting selection from phenotype-based to DNA-based methods. This shift offers several advantages, including cost-effectiveness, time efficiency, and improved selection precision. Molecular selection can be applied to seedlings and allows for enriching populations with heterozygous individuals using codominant markers (Cobb *et al.* 2019).

Various resources, such as QTLs, cloned genes, closely related markers, and donor lines, have become available for multiple crops. These resources facilitate their transfer to elite lines through MAS. For example, efforts have been made to enhance resistance to diseases like rice blast and bacterial blight in rice, as well as improve resistance in crops like chili, brassica, and wheat.

When the donor line originates from a wild relative, a common challenge is the occurrence of linkage drag, where undesirable alleles from physically linked loci may be introduced along with the desired gene. To mitigate this issue, background selection is employed, involving the use of markers to monitor the percentage of the recurrent parent genome. However, when recombination is limited, alternative strategies, such as using mutants that restore recombination, are applied to reduce the introgressed region.

One of the notable advantages of MAS is its ability to expedite the improvement of crop lines, particularly when combined with early-stage testing. Techniques like "speed breeding" enable the achievement of multiple generations per year for various crops, significantly accelerating the breeding process.

Comparative studies have evaluated the effectiveness of MAS against other selection methods, with outcomes dependent on factors like the genetic basis of the targeted trait. While MAS is effective in many cases, it is often integrated with other approaches in breeding programs to optimize results, aligning with the specific genetic control of the trait of interest. For instance, in hybrid rice production, MAS is combined with selection for hybrid genes related to male sterility, resulting in disease-resistant hybrid rice lines.

12.6.3 Genomic Selection and Machine Learning

Plant breeding programs are traditionally time-consuming and resource-intensive, often taking several years to release new crop varieties. This extended timeline is due to the need for breeders to generate genetic diversity through crossbreeding, stabilize the resulting lines, and meticulously evaluate candidate varieties across various environmental conditions to select those with the best agronomic performance. In the era before genomics, traditional breeding strategies relied heavily on quantitative genetic approaches that incorporated pedigree information to estimate Best Linear Unbiased Predictors (BLUPs) (Henderson *et al.* 1976).

The landscape of plant breeding underwent a significant transformation in the 1990s with the advent of molecular genetics techniques. These techniques introduced a plethora of molecular markers, significantly expanding the toolbox available to plant breeders. More than 10,000 Quantitative Trait Loci (QTLs) have been identified across 12 plant species in over 120 studies, all aimed at improving economically important quantitative traits.

Over the past few decades, molecular markers have been seamlessly integrated into traditional breeding methods through Marker-Assisted Selection (MAS). This approach involves selecting individuals based on markers associated with specific QTLs that have major effects. However, many economically important quantitative traits are controlled by numerous genes with minor effects, making them challenging to effectively utilize in breeding. Identifying these small-effect loci has been challenging, and even when successful, multiple QTLs are often present, complicating their simultaneous use in breeding. Additionally, MAS has encountered limitations when addressing complex quantitative traits due to the instability of QTLs across various environments, influenced by QTL-environment interactions (Bernardo *et al.* 2016).

The emergence of high-density Single Nucleotide Polymorphism (SNP) arrays paved the way for the implementation of Genomic Selection (GS). GS aims to circumvent the limitations of MAS by assessing an individual's genetic potential rather than precisely locating specific QTLs. Various statistical models for GS have been developed, each accommodating different assumptions about the distribution of SNP effects.

GS hinges on the presence of a training population and a test population, both consisting of individuals from the reference population with phenotypic data pertaining to the target traits and genotypic data spanning their entire genomes. A mathematical equation, accounting for marker effects estimated from the training population, is utilized to compute the Genomic Estimated Breeding Value (GEBV) for an individual. Cross-validation is commonly employed to evaluate the model's predictive accuracy. GS has demonstrated its effectiveness in numerous crop species, enhancing selection precision and the rate of genetic improvement (Xu *et al.* 2020).

From the implementation of GS in crop breeding programs, three fundamental insights have emerged. First, the training population must be substantial and encompass individuals closely related to the selection candidates. Second, regular updates to the reference population, encompassing both training and test populations, are essential to sustain the accuracy of GEBV over time. Lastly, GS is the preferred approach over MAS when dealing with quantitative resistance that relies on multiple minor genes.

With the progress of GS, there has been a substantial increase in available data, fostering interdisciplinary research collaborations, including those with computer science and machine learning. Machine Learning (ML), a subset of Artificial Intelligence (AI), has gained prominence in managing extensive datasets. ML includes supervised, unsupervised, and reinforcement learning techniques. ML models prioritize achieving accurate predictions by learning patterns within data, making them well-suited for handling complex genetic information.

Among ML methods, Support Vector Machine (SVM) and Random Forest have demonstrated effectiveness in analyzing genetic data, particularly in studies related to disease resistance. Artificial Neural Networks (ANNs) represent a special category of ML, inspired by biological neural networks. ANNs comprise interconnected neurons, and deep learning (DL) entails networks with multiple hidden layers. DL finds applications in genotype-phenotype correlation, causal variant discovery, and image-based phenotyping in plants.

DL is particularly advantageous when dealing with high-throughput phenotyping data and environmental factors, often referred to as "envirotyping." Advances in high-throughput phenotyping enable researchers to collect environmental data through various methods, further enhancing the analysis of genotype-environment interactions and the prediction of phenotypic traits (Citri *et al.* 2008 and Frank *et al.* 2019).

12.6.4. Transcriptomics

Transcriptomics is a powerful molecular biology technique used to study gene expression patterns within an organism, tissue, or cell under different conditions, including during pathogen interactions. When applied to pathogen-host interactions, transcriptomics provides valuable insights into how the host responds to the presence of a pathogen and how the pathogen manipulates the host's gene expression to its advantage. Here's an overview of the use of transcriptomics in understanding gene expression during pathogen interactions:

- ❒ **Sample Collection and RNA Isolation**: The process begins with the collection of biological samples from the host organism or tissue of interest, typically at different time points during the infection process. These samples are then used to isolate RNA, which represents the transcriptome of the cells at a given moment.

- ❒ **Library Preparation**: Isolated RNA is converted into complementary DNA (cDNA) libraries. This step involves reverse transcription of RNA into cDNA, followed by library preparation, which usually includes fragmentation, adapter ligation, and amplification.

- ❒ **Sequencing**: Next-generation sequencing (NGS) technologies, such as RNA-Seq, are employed to sequence the cDNA libraries. This step provides a snapshot of all the RNA molecules present in the sample, including mRNAs, non-coding RNAs, and small RNAs.

- ❒ **Data Analysis**:
  - ☆ **Alignment**: Sequencing reads are aligned to a reference genome or transcriptome to determine the origin of each read. This step helps in identifying the genes from which the RNA transcripts originate.

  - ☆ **Quantification**: The abundance of each RNA transcript is quantified by counting the number of sequencing reads that map to specific genes. This quantification is usually represented as Fragments per Kilobase of transcript per Million mapped reads (FPKM) or Transcripts Per Million (TPM).

  - ☆ **Differential Expression Analysis**: Statistical methods are applied to compare gene expression levels between different conditions, such as infected vs. uninfected samples or different time points post-infection. Genes that show significant changes in expression are identified.

☆ **Functional Annotation**: Bioinformatic tools are used to annotate the differentially expressed genes, categorize them based on their functions, and predict the biological processes they are involved in.

☐ **Visualization**: Data visualization tools, such as heatmaps, volcano plots, and pathway analysis, help researchers interpret the results and identify patterns of gene expression that are associated with the host response to the pathogen.

☐ **Validation**: Transcriptomic findings are often validated using other experimental techniques, such as quantitative real-time PCR (qPCR) or immunoblotting, to confirm changes in gene expression at the protein level.

Transcriptomics can provide several key insights into pathogen-host interactions:

☐ **Identification of Key Players**: Researchers can identify genes that are upregulated or downregulated during infection, helping to pinpoint key players in the host response.

☐ **Temporal Dynamics**: Transcriptomics can reveal the temporal dynamics of gene expression changes over the course of infection, shedding light on the kinetics of the host response.

☐ **Pathways and Processes**: It helps uncover the biological pathways and processes that are affected by the infection, including immune responses, signal transduction, and metabolic changes.

☐ **Host-Pathogen Interactions**: By studying both host and pathogen transcripts simultaneously, transcriptomics can provide insights into the molecular dialogue between the host and the pathogen, including the strategies employed by the pathogen to evade the host's defense mechanisms.

☐ **Biomarker Discovery**: Transcriptomic data can lead to the discovery of potential diagnostic or prognostic biomarkers for diseases caused by pathogens.

In summary, transcriptomics is a valuable tool for unraveling the intricacies of gene expression during pathogen interactions. It helps researchers gain a comprehensive understanding of how pathogens manipulate host gene expression and how the host responds, ultimately contributing to our knowledge of infectious diseases and the development of novel therapeutic strategies.

12.6.5. Effectoromics

Effectoromics is an innovative approach among "omics" technologies that plays a crucial role in understanding and managing pathogenesis. It focuses on the identification of candidate pathogen effectors (Avr) and utilizes them as functional markers to define corresponding host resistance (R) and susceptibility (S) genes. Effectoromics has evolved significantly over time, involving methods such as physical isolation, in silico predictions, functional characterization of effectors, and identification of host targets. Here are some key points related to effectoromics:

- ❑ **Evolution of Effector Discovery**: Effector discovery techniques have advanced from physical isolation and computational predictions to functional characterization of plant pathogen effectors and the identification of their host targets. These methods have greatly improved our understanding of host-pathogen interactions.

- ❑ **Examples in Crops**: While effectoromics has shown promise, there are relatively few examples reported in crop research. For instance, in the case of tomato late blight caused by Phytophthora infestans, several genes for field partial resistance (Rpi-blb1, Rpi-blb2, and Rpi-vnt1.1) have been discovered, along with the corresponding avirulence genes (Avrblb1, Avrblb2, and Avrvnt1). These discoveries have provided insights into resistance mechanisms (Champouret *et al*. 2009, Chen *et al*. 2017 and Roman *et al*. 2017).

- ❑ **New Avenues**: Recent research has identified a new family of Phytophthora small extracellular cysteine-rich proteins (PcF/SCR) recognized by solanaceous plants. When combined with known NLR genes, these effectors offer the potential to target a wide spectrum of Phytophthora infestans strains, which could be valuable for potato breeding (Lin *et al*. 2020).

- ❑ **Wheat Pathogens**: Effectoromics has also made progress in identifying candidate effectors from wheat pathogens. Some of these effectors have been functionally validated and cloned. For instance, susceptibility to tan spot caused by Pyrenophora tritici-repentis is correlated with sensitivity to the effector ToxA. A high-throughput screening method using ToxA infiltration has been developed to identify wheat genotypes carrying the susceptible allele at the Tsn1 locus (Ramachandran *et al*. 2017 and Hewitt *et al*. 2021).

- ❑ **Challenges in Quantitative Resistance**: Diseases like Fusarium head blight (FHB) or *Blumeria graminis* f. sp. hordei pose challenges as quantitative resistance involves numerous effectors. More research is needed to discover the full effector repertoire of pathogens like *F. graminearum* (Gorash *et al*. 2021 and Pedersen *et al*. 2012).

❏ **Database Resources**: Effectoromics research benefits from databases like the Pathogen–Host Interactions database (PHI-base), which contains curated information on genes from various pathogens and hosts. These databases are instrumental in enhancing our knowledge of effector interactions (Urban *et al.* 2020).

❏ **Future Directions**: Effectoromics holds great potential, but there is a need for more research in crop systems. Advances in machine learning and deep learning, combined with effector and genome sequence databases, may enable the prediction of genes that interact with effectors to confer resistance or susceptibility in crops.

12.6.6 New Breeding Technologies (NBTs)

Genetic engineering presents significant prospects for enhancing resistance against pathogens in crops. Initially, the predominant approach involved the introduction of foreign genes (transgenic) into numerous crop varieties, resulting in the commercial approval of genetically modified food crops. However, this method, in conjunction with traditional breeding and marker-assisted selection (MAS), encountered regulatory and sustainability challenges. More recently, there has been a shift towards direct manipulation of plant genomes. This can be accomplished by inserting natural resistance genes using genetic transformation or employing environmentally friendly new breeding technologies (NBTs) like gene modification and editing. Notably, approaches such as cisgenesis, intragenesis, and genome editing (GE) have gained widespread adoption. Additionally, researchers have explored the use of RNA interference and transgrafting to enhance disease resistance in woody fruit species. These advances hold substantial potential for fortifying crop resilience against pathogenic threats.

12.6.6.1 Cisgenesis/Intragenesis

The utilization of a species' genetic pool or that of a closely related species, akin to what is available in traditional breeding, can be harnessed for genetic modifications in crop plants through techniques like cisgenesis or intragenesis (Schouten *et al.* 2006 and Holme *et al.* 2013). This entails either transferring resistance genes from related species or amplifying the expression of genes already existing in the crop itself. This approach offers significant time savings and avoids the issue of linkage drag, which is associated with traditional gene transfer through crossbreeding. The success of this method relies on the isolation of complete functional genes, along with their associated regulatory sequences like promoters and terminators. Advances in sequencing technologies and the accessibility of extensive genome information have facilitated this process. However, it's worth noting that this approach is applicable to only a limited number of crop species. For instance, in cereal crops, cisgenesis has been employed to enhance pathogen resistance, primarily in wheat. In contrast, horticultural, fruit, and ornamental crops, many of which are heterozygous and propagated vegetatively, have seen successful improvements in disease resistance through cisgenesis and intragenesis techniques (Singh *et al.* 2018).

In the case of potatoes, the Durable Resistance against Phytophthora (DuRPh) program was initiated, aiming to introduce multiple (RERR)-genes from wild potatoes, along with their native regulatory sequences, into susceptible cultivated potato varieties. This was achieved through a cisgenic approach, and their performance was assessed through field trials (Holme *et al.* 2013 and Haverkort *et al.* 2016). Furthermore, marker-free Agrobacterium-mediated transformation led to the development of cisgenic potato plants that stacked blight resistance R-genes from *Solanum stoloniferum* and *S. venturi*, resulting in broad-spectrum resistance to late blight without altering the original variety's characteristics (Jo *et al.* 2014). Other examples of successful application of cisgenic and intragenic approaches include melon plants with enhanced resistance against powdery mildew, strawberry plants resistant to gray mold, and woody fruit species like apples and grapes engineered for disease resistance. Grapes, for instance, exhibited reduced powdery mildew development and black rot severity in field tests.

To enhance the application of cisgenesis and intragenesis, measures need to be implemented to address specific challenges, such as gene expression variability or silencing of endogenous genes based on the insertion position of the cis/intragene or the presence of extraneous sequences. Nevertheless, cisgenic lines could serve as valuable starting points in selecting parents for breeding programs aimed at enhancing disease resistance.

12.6.6.2. Genome Editing (GE)

Genome editing (GE) is a sophisticated genetic modification technique that enables precise and efficient modification of genomic sequences using engineered nucleases. These nucleases include zinc finger nucleases (ZFNs), transcription activator-like effector nucleases (TALENs), and more recently, clustered regularly interspaced short palindromic repeat (CRISPR) systems with CRISPR-associated protein 9 (Cas9) (CRISPR/Cas9) (Kamburova *et al.* 2017). These nucleases induce double-strand breaks (DSBs) in specific genome regions, allowing for the deletion, replacement, or insertion of targeted sequences. DNA repair mechanisms then make these modifications useful for gene knockout or insertion, creating new and improved alleles for specific traits while replacing unfavorable ones. This approach relies on knowledge of genomic sequences, which are now available for major crop plants.

Unlike ZFNs and TALENs, which use protein motifs for target identification, the CRISPR/Cas9 system relies on base complementarity between CRISPR RNA and target DNA or RNA molecules for cleavage specificity. It offers several advantages, including simplicity, high efficiency, precision, ease of use in the laboratory, broad applicability, and the ability to target multiple genes simultaneously (multiplexing). Additionally, CRISPR/Cas9 is cost-effective, and vector design is relatively straightforward thanks to improved bioinformatics tools. Designing multiple targets for one gene can enhance the probability of obtaining homozygous mutations in the first generation, a particularly important consideration for woody plants with longer generation times. Notably, CRISPR/Cas9 can target not only open reading frame (ORF) regions but also cis-regulatory elements like promoters.

CRISPR/Cas9 can achieve efficient, transgene-free genome editing in plants through various methods, including *Agrobacterium tumefaciens* transformation, protoplast transformation, or direct delivery of guide RNA (gRNA) and Cas9 to plant cells. Most modifications aimed at enhancing pathogen resistance have been conducted using Agrobacterium-mediated transformation of different plant tissues.

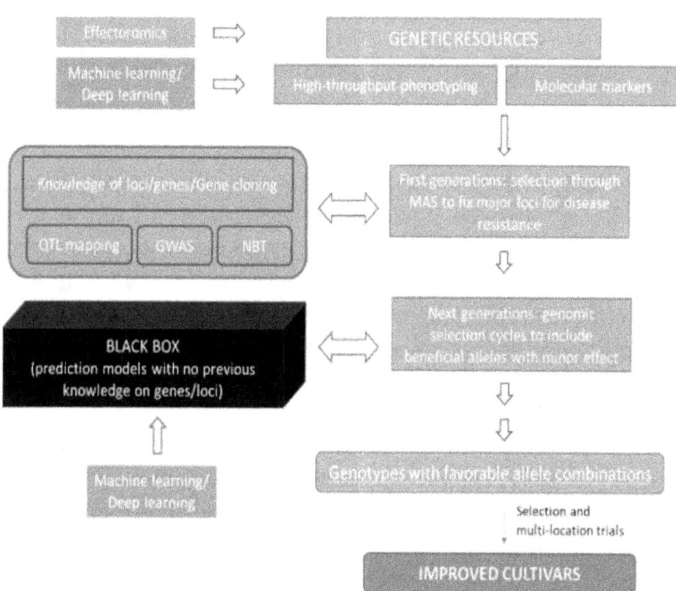

Figure 2. Workflow by integrating the different molecular approaches to obtain improved cultivars for disease resistance in crops (Mores et al. 2021).

GE has been successfully applied to confer resistance against various pathogens, including bacteria, fungi, and DNA and RNA viruses. For instance, in rice, GE targeted genes involved in the response to *Xanthomonas oryzae pv. oryzae (Xoo)* infection, a significant bacterial blight disease. Similar approaches have been used in tomato and citrus to enhance resistance against multiple plant pathogens. In woody fruit species like apple and grapes, GE has been employed to increase resistance to fire blight and other diseases (Malnoy *et al.* 2016).

Regarding fungal pathogens, GE has been applied to create resistance against powdery mildew (PM) in wheat by targeting the Mlo gene (Shan *et al.*

2013). Fruit species like tomatoes have seen increased resistance to bacterial speck disease through GE targeting of the SlJAZ2 gene.

GE has also been utilized to improve resistance against oomycetes, such as *Phytophthora sojae*, by editing effector genes like Avr4/6, preventing recognition by plant R proteins. Moreover, GE has been effective against both DNA and RNA plant viruses, using sgRNAs to target viral genetic elements and prevent replication. Another approach involves targeting host factors necessary for viral infection, such as eIF4E, to enhance resistance against RNA-based viruses (Chandrasekaran *et al.* 2016).

However, challenges remain, including potential impacts on crop growth, off-target effects, and the evolution of viral mutations to escape CRISPR/ Cas9 cleavage (Zhao *et al.* 2020). Strategies to mitigate these challenges include targeting essential genomic regions for viral replication or movement, multiplexing guide RNAs, and using Cas12a (Cpf1) to reduce the likelihood of escape viral variants. Edited plants with modified resistance genes can be valuable for integration into breeding programs aimed at rapidly obtaining lines with favorable allelic combinations and improved resistance.

12.7 Challenges and Future Directions

Genomic approaches have transformed crop breeding by offering potent tools to develop disease-resistant varieties, yet they are accompanied by significant challenges. Disease resistance traits are often polygenic and subject to intricate gene interactions, complicating their identification. Limited genetic diversity in certain crop varieties and rapid pathogen evolution pose additional hurdles. Ethical concerns surrounding genetically modified organisms (GMOs) and issues related to intellectual property rights and resource accessibility further complicate matters. Functional validation, resource constraints, lengthy breeding cycles, and regulatory hurdles are practical obstacles. Environmental, ecological, and equity considerations also loom large, underscoring the need for balanced approaches that integrate genomics with traditional breeding methods. Overcoming these challenges necessitates collaborative efforts to ensure the widespread success of genomic approaches in breeding disease-resistant crops.

Ethical and regulatory considerations in plant gene editing are paramount. They encompass environmental impacts, biodiversity conservation, intellectual property rights, food safety, and labeling. Cross-border movement, public engagement, and responsible innovation require robust regulations and global collaboration. Ensuring containment measures and coexistence with existing farming practices are vital for agricultural diversity. Balancing innovation with environmental and ethical concerns necessitates harmonized global approaches.

The future of plant genomics and disease resistance breeding presents a compelling landscape of opportunities. Advancements in precision genomic editing technologies, coupled with a deeper understanding of gene function through functional genomics, will empower scientists to develop highly specific and durable disease-resistant crops. Integration of bioinformatics and big data analytics will accelerate gene discovery and enhance our ability to predict crop-pathogen interactions. Genomic selection and multi-omics integration will streamline breeding processes, while synthetic biology may enable the creation of custom resistance genes. Advanced breeding techniques and eco-friendly disease control strategies will promote sustainable agriculture, especially in the face of climate change. Global collaboration, ethical considerations, and robust regulatory frameworks will be essential to harness these innovations for the benefit of global food security and the environment. In the coming years, the field will continue to evolve; pushing the boundaries of what is possible in plant genomics and disease resistance breeding.

12.8 Conclusions and Prospects for New Breeding Scenarios

The rising occurrence and severity of diseases due to global climate changes are impacting the world's crop production. To address this challenge and expedite the development of disease-resistant crop varieties, genomics-assisted breeding has emerged as a pivotal tool in breeding programs. A progressive breeding strategy for enhancing resistance in herbaceous plants, as depicted in Figure 2, commences with the selection of parent lines recognized for their robust disease resistance, which includes both prominent and less-defined resistance alleles. Lines produced through genome editing of resistance genes (R genes) are also valuable assets to this initiative. Furthermore, potential candidate lines can be pinpointed by screening for effector targets of resistance genes; those exhibiting a favorable response to the effector hold promise for breeding superior varieties.

Molecular markers tied to alleles at significant loci or precise markers crafted based on the gene sequences of cloned R genes can be utilized in marker-assisted selection (MAS). This expedites the rapid establishment of desired allele combinations in the early phases of the breeding program. Subsequently, several cycles of genomic selection can be incorporated to identify minor quantitative resistance genes, which contribute to more durable resistance but pose challenges in terms of identification and selection through traditional means.

Machine learning, especially in managing extensive datasets, can be seamlessly integrated into this breeding scheme to complement MAS and genomic selection. It aids in data acquisition, the interpretation of phenotypic data, and the forecasting of breeding values. This comprehensive integration of various methods significantly trims the duration of the breeding process and reduces selection costs, ultimately fostering the development of crop varieties boasting enhanced disease resistance.

References

Abe, A., Kosugi, S., Yoshida, K., Natsume, S., Takagi, H., Kanzaki, H., Matsumura, H., Yoshida, K., Mitsuoka, C., Tamiru, M., *et al.* (2012). Genome Sequencing Reveals Agronomically Important Loci in Rice Using MutMap. *Nature Biotechnology, 30*(2), 174–178.

Bernardo, R. (2016). Bandwagons I, Too, Have Known. *Theoretical and Applied Genetics, 129*(12), 2323–2332.

Cai, J., Wang, S., Su, Z., Li, T., Zhang, X., & Bai, G. (2019). Meta-analysis of QTL for *Fusarium* Head Blight Resistance in Chinese Wheat Landraces. *Crop Journal, 7*(6), 784–798.

Champouret, N., Bouwmeester, K., Rietman, H., van der Lee, T., Maliepaard, C., Heupink, A., van de Vondervoort, P.J.I., Jacobsen, E., Visser, R.G.F., van der Vossen, E.A.G., *et al.* (2009). *Phytophthora Infestans* Isolates Lacking Class I ipiO Variants are Virulent on Rpi-blb1 Potato. *Molecular Plant-Microbe Interactions, 22*(12), 1535–1545.

Chandrasekaran, J., Brumin, M., Wolf, D., Leibman, D., Klap, C., Pearlsman, M., Sherman, A., Arazi, T., & Gal-On, A. (2016). Development of Broad Virus Resistance in Non-transgenic Cucumber Using CRISPR/Cas9 Technology. *Molecular Plant Pathology, 17*(7), 1140–1153.

Chen, Y., & Halterman, D.A. (2017). *Phytophthora Infestans* Effectors IPI-O1 and IPI-O4 Each Contribute to Pathogen Virulence. *Phytopathology, 107*(5), 600–606.

Citri, A., & Malenka, R.C. (2008). Synaptic Plasticity: Multiple Forms, Functions, and Mechanisms. *Neuropsychopharmacology Reviews, 33*(1), 18–41.

Cobb, J.N., Biswas, P.S., & Platten, J.D. (2019). Back to the Future: Revisiting MAS as a Tool for Modern Plant Breeding. *Theoretical and Applied Genetics, 132*(3), 647–667.

Dracatos, P.M., Bartos, J., Elmansour, H., Singh, D., Karafiatova, M., & Zhang, P. (2019). The Coiled-coil NLR Rph1, Confers Leaf Rust Resistance in Barley Cultivar Sudan. *Plant Physiology, 179*(3), 1362–1372.

Frank, M.G. (2019). The Role of Glia in Sleep-Wake Regulation and Function. In *Handbook of Behavioral Neuroscience* (Vol. 30, pp. 195–204). Elsevier.

Gorash, A., Armonien e, R., & Kazan, K. (2021). Can Effectoromics and Loss-of-susceptibility be Exploited for Improving Fusarium Head Blight Resistance in Wheat? *Crop Journal, 9*, 1–16.

Haverkort, A.J., Boonekamp, P.M., Hutten, R., Jacobsen, E., Lotz, L.A.P., Kessel, G.J.T., Vossen, J.H., & Visser, R.G.F. (2016). Durable Late Blight Resistance in Potato through Dynamic Varieties Obtained by Cisgenesis: Scientific and Societal Advances in the DuRPh Project. *Potato Research, 59*(1), 35–66.

Henderson, C.R., & Quaas, R.L. (1976). Multiple Trait Evaluation Using Relative's Records. *Journal of Animal Science, 43*(1), 11–88.

Hewitt, T., Muller, M.C., Molnar, I., Mascher, M., Holusova, K., Simkova, H., Kunz, L., Zhang, J., Li, J., Bhatt, D., *et al.* (2021). A Highly Differentiated Region of Wheat Chromosome 7AL encodes a Pm1a Immune Receptor That Recognizes Its Corresponding AvrPm1a Effector from *Blumeria graminis*. *New Phytologist, 229*(6), 2812–2826.

Hickey, L.T., Germán, S.E., Pereyra, S.A., Diaz, J.E., Ziems, L.A., Fowler, R.A., Platz, G.J., Franckowiak, J.D., & Dieters, M.J. (2017). Speed Breeding for Multiple Disease Resistance in Barley. *Euphytica, 213*(3), 64.

Holme, I.B., Wendt, T., & Holme, P.B. (2013). Intragenesis and Cisgenesis as Alternatives to Transgenic Crop Development. *Plant Biotechnology Journal, 11*(4), 395–407.

Jo, K.R., Kim, C.J., Kim, S.J., Kim, T.Y., Bergervoet, M., Jongsma, M.A., Visser, R.G.F., Evert Jacobsen, E., & Vossen, J.H. (2014). Development of Late Blight Resistant Potatoes by Cisgene Stacking. *BMC Biotechnology, 14*, 50.

Kamburova, V.S., Nikitina, E.V., Shermatov, S.E., Buriev, Z.T., Kumpatla, S.P., Emani, C., & Abdurakhmonov, I.Y. (2017). Genome Editing in Plants: An Overview of Tools and Applications. *International Journal of Agronomy, 2017*, 7315351.

Li, W., Deng, Y., Ning, Y., He, Z., & Wang, G.L. (2020). Exploiting Broad-spectrum Disease Resistance in Crops: From Molecular Dissection to Breeding. *Annual Review of Plant Biology, 71*, 575–603.

Lin, X., Wang, S., de Rond, L., Bertolin, N., Wouters, R.H.M., Wouters, D., Domazakis, E., Bitew, M.K., Win, J., Dong, S., *et al.* (2020). Divergent Evolution of PcF/SCR74 Effectors in Oomycetes is Associated with Distinct Recognition Patterns in *Solanaceous* Plants. *mBio, 11*, e00947-20.

Mahlein, A.K., Kuska, M.T., Thomas, S., Wahabzada, M., Behmann, J., Rascher, U., & Kersting, K. (2019). Quantitative and Qualitative Phenotyping of Disease Resistance of Crops by Hyperspectral Sensors: Seamless Interlocking of Phytopathology, Sensors, and Machine Learning is Needed! *Current Opinion in Plant Biology, 50*, 156–162.

Malnoy, M., Viola, R., Jung, M.H., Koo, O.J., Kim, S., Kim, J.S., Velasco, R., & Nagamangala Kanchiswamy, C. (2016). DNA-free Genetically Edited Grapevine and Apple Protoplast Using CRISPR/Cas9 Ribonucleoproteins. *Frontiers in Plant Science, 7*, 1904.

Mores, A., Borrelli, G.M., Laidò, G., Petruzzino, G., Pecchioni, N., Amoroso, L.G.M., Desiderio, F., Mazzucotelli, E., Mastrangelo, A.M., & Marone, D. (2021). Genomic Approaches to Identify Molecular Bases of Crop Resistance to Diseases and to Develop Future Breeding Strategies. *International Journal of Molecular Sciences, 22*, 5423.

Niks, R.E., Qi, X., & Marcel, T.C. (2015). Quantitative Resistance to Biotrophic Filamentous Plant Pathogens: Concepts, Misconceptions, and Mechanisms. *Annual Review of Phytopathology, 53*, 445–470.

Pedersen, C., Ver Loren van Themaat, E., McGuffin, L.J., Abbott, J.C., Burgis, T.A., Barton, G., Bindschedler, L.V., Lu, X., Maekawa, T., Wessling, R., *et al.* (2012). Structure and Evolution of Barley Powdery Mildew Effector Candidates. *BMC Genomics, 13*, 694.

Pilet-Nayel, M.L., Moury, B., Caffier, V., Montarry, J., Kerlan, M.C., Fournet, S., Durel, C.E., & Delourme, R. (2017). Quantitative Resistance to Plant Pathogens in Pyramiding Strategies for Durable Crop Protection. *Frontiers in Plant Science, 8*, 1838.

Piquerez, S.J., Harvey, S.E., Beynon, J.L., & Ntoukakis, V. (2014). Improving Crop Disease Resistance: Lessons from Research on Arabidopsis and Tomato. *Frontiers in Plant Science, 5*, 671.

Ramachandran, S.R., Yin, C., Kud, J., Tannaka, K., Mahoney, A.K., Xiao, F., & Hilbert, S.H. (2017). Effectors from Wheat Rust Fungi Suppress Multiple Plant Defense Responses. *Phytopathology, 107*(1), 75–83.

Ramirez-Gonzalez, R.H., Segovia, V., Bird, N., Fenwick, P., Holdgate, S., Berry, S., Jack, P., Caccamo, M., & Uauy, C. (2015). RNA-Seq Bulked Segregant Analysis Enables the Identification of High-resolution Genetic Markers for Breeding in Hexaploid Wheat. *Plant Biotechnology Journal, 13*(5), 613–624.

Roman, M.L., Izarra, M., Lindqvist-Kreuze, H., Rivera, C., Gamboa, S., Tovar, J.C., Forbes, G., Kreuze, J.F., & Ghislain, M. (2017). R/Avr Gene Expression Study of Rpi-vnt1.1 Transgenic Potato Resistant to the *Phytophthora infestans* Clonal Lineage EC-1. *Plant Cell Tissue and Organ Culture (PCTOC), 131*(2), 259–268.

Sánchez-Martín, J., Steuernagel, B., Ghosh, S., Herren, G., Hurni, S., Adamski, N., Vrána, J., Kubaláková, M., Krattinger, S.G., Wicker, T., *et al.* (2016). Rapid Gene Isolation in Barley and Wheat by Mutant Chromosome Sequencing. *Genome Biology, 17*, 221.

Schouten, H.J., Krens, F.A., & Jacobsen, E. (2006). Cisgenic Plants are Similar to Traditionally Bred Plants: International Regulations for Genetically Modified Organisms Should be Altered to Exempt Cisgenesis. *EMBO Reports, 7*(8), 750–753.

Shan, Q., Wang, Y., & Li, J. (2013). Targeted Genome Modification of Crop Plants Using a CRISPR–Cas System. *Nature Biotechnology, 31*(8), 686–688.

Singh, V., Singh, S., Shikha, K., & Kumar, A. (2018). Cisgenesis a Sustainable Approach of Gene Introgression and Its Utilization in Horticultural Crops: A Review. *International Journal of Current Microbiology and Applied Sciences, 7*(8), 5002–5009.

St. Clair, D.A. (2010). Quantitative Disease Resistance and Quantitative Resistance Loci in Breeding. *Annual Review of Phytopathology, 48*, 247–268.

Steuernagel, B., Periyannan, S.K., Hernández-Pinzón, I., Witek, K., Rouse, M.N., Yu, G., Hatta, A., Ayliffe, M., Bariana, H., Jones, J.D., *et al.* (2016). Rapid Cloning of Disease-resistance Genes in Plants Using Mutagenesis and Sequence Capture. *Nature Biotechnology, 34*(6), 652–655.

Thind, A.K., Wicker, T., Simkova, H., Fossati, D., Moullet, O., Brabant, C., Vrana, J., Dolezel, J., & Krattinger, S.G. (2017). Rapid Cloning of Genes in Hexaploid Wheat Using Cultivar-specific Long-range Chromosome Assembly. *Nature Biotechnology, 35*(9), 793–796.

Thomson, M.J. (2014). High-throughput SNP Genotyping to Accelerate Crop Improvement. *Plant Breeding and Biotechnology, 2*(3), 195–212.

Urban, M., Cuzick, A., Seager, J., Wood, V., Rutherford, K., Venkatesh, S.Y., De Silva, N., Martinez, M.C., Pedro, H., Yates, A.D., *et al.* (2020). PHI-base: The Pathogen–host Interactions Database. *Nucleic Acids Research, 48*(D1), D613–D620.

Varshney, R.K., Bansal, K.C., Aggarwal, P.K., Datta, S.K., & Craufurd, P.Q. (2011). Agricultural Biotechnology for Crop Improvement in a Variable Climate: Hope or Hype? *Trends in Plant Science, 16*(7), 363–371.

Xu, Y., Liu, X., Fu, J., Wang, H., Wang, J., Huang, C., Prasanna, B.M., Olsen, M.S., Wang, G., & Zhang, A. (2020). Enhancing Genetic Gain through Genomic Selection: From Livestock to Plants. *Plant Communications, 1*(1), 1.

Zhao, Y., Yang, X., Zhou, G., & Zhang, T. (2020). Engineering Plant Virus Resistance: From RNA Silencing to Genome Editing Strategies. *Plant Biotechnology Journal, 18*(2), 328–336.

Chapter–13

Breeding for Insect Pest Resistance Using Genomic Tools

Sravani Devagudi[1]**, Pavani Narigapalli**[2]**, Manoj Kumar S. C.**[3] **and Divya Ponnangi**[4]

123 Ph.D. Scholars, Department of Genetics and Plant Breeding

4   Ph. D Scholar, Department of Entomology, S.V. Agricultural College- 517502, Acharya N.G. Ranga Agricultural University

**Corresponding author: sravanidevagudi20@gmail.com*

13.1 Introduction

Enhancing crop yield is a crucial issue to tackle, especially considering the current challenges posed by rapid population growth and unpredictable climatic changes. In the 21st century, agriculture faces continuous obstacles arising from various biological and environmental stressors, with insect pests standing out as a significant concern. These insects contribute to diminishing potential harvests, leading to stagnant agricultural output. Despite operating control measures, economically vital crops still experience substantial yield reductions. Farmers, nevertheless, have turned to chemical pesticides as a solution for managing insect pests. However, the extensive application of these pesticides can pose risks to both human health and the environment. These challenges have motivated researchers worldwide to create innovative and ecologically conscious methods for insect control. As a result, modern agriculture places a central emphasis on attaining higher yields using the available land and resources, all in the pursuit of global food security and sustainable agricultural practices.

Genomic tools encompass a range of advanced techniques that involves the comprehensive understanding of the genetic information contained within the genomes of both crops and insects. The integration of genomics into pest resistance breeding has unlocked avenues for targeted genetic manipulations that can enhance the innate defenses of crops against pest attacks. This approach not only expedites the breeding process but also provides the opportunity to develop crop varieties with durable resistance traits.

The cornerstone of pest management lies in host plant resistance, representing one of the most esteemed strategies in advanced agriculture (Horgan *et al.*, 2020 and El-Dessouki *et al.*, 2022). This is due to inherent heritable traits within plants, endowing them with the ability to endure less harm compared to their less-endowed counterparts. Crop varieties resistant to insects decrease the insect pest population by enhancing their capacity to endure damage. The interaction between the insect and the plant is determined by three distinct forms of resistance: antibiosis, antixenosis (non-preference), and tolerance presented in Fig.1.

13.2 Methods of Pest Resistance

13.2.1. Antibiosis

This method involves the creation of genetically modified plants that negatively affect the development, survival, or reproduction of insect pests that feed on them. These plants release compounds that are toxic to the pests, impairing their growth and overall fitness.

13.2.2. Antixenosis

Antixenosis, also known as non-preference or deterrence, focuses on making plants less attractive to insects. Through genetic modifications, plants can be altered to produce compounds that repel or deter insects from feeding on them, making them choose alternative food sources.

13.2.3. Non-Tolerance

Non-tolerance involves developing crop varieties that can better withstand insect damage without showing significant negative effects on yield. This is achieved through a combination of genetic traits that enable plants to recover from insect feeding or damage more efficiently.

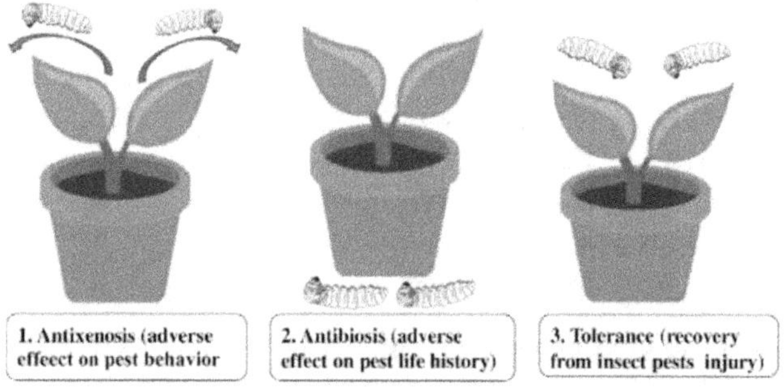

Fig.13.1. Mechanisms of Host Plant Resistance

In 1782, the journey towards creating plants resistant to insects commenced when Havens authored a paper on a type of wheat cultivar resistant to the Hessian fly. Since that era, numerous insect-resistant cultivars have been successfully developed by global and national research institutions, as well as private enterprises, employing both traditional and biotechnological methods (Jaiswal *et al.*, 2018). Over the past three decades, substantial progress has been achieved in comprehending the fundamental biochemical principles that underscore such resistance, along with the identification of the genes involved (Joshi *et al.*, 2020). Furthermore, in the pursuit of global food security and sustainable agriculture, the central aim of contemporary farming revolves around augmenting yields while making optimal use of available land and resources. As a consequence, it becomes imperative to harness forward-looking technologies to effectively manage pests and ensure ample food supply in the coming years. Diverse innovations in plant protection have been devised to regulate, prevent, and oversee these pests. There's a prevailing inclination toward adopting novel and advanced biotechnological strategies, recognized for their remarkable efficiency and swift results when contrasted with traditional methods. These methods have the potential to form the cornerstone of safeguarding crops against a wide array of insect pests. In light of these considerations, modern biotechnology emerges as a compelling avenue, presenting promising alternatives and capitalizing on the potential offered by these novel biotechnological tools to develop various crop plants that exhibit resilience against a diverse spectrum of insect pests (Kumari *et al.*, 2022).

13.3. Biotechnological Approaches for Insect Pest Management

Biotechnological approaches for insect pest management involve the use of genetic, molecular, and cellular techniques to develop strategies that effectively control insect pests while minimizing the negative impacts on non-target organisms and the environment. These approaches often provide more specific and environmentally friendly alternatives to traditional chemical pesticides.

13.3.1. Gene Transformation

Gene transformation, also known as genetic transformation, refers to the process of introducing foreign DNA into the genome of an organism. This technology allows scientists to modify an organism's genetic makeup by adding new genes or altering existing ones. Gene transformation is a key technique in various fields, including biotechnology, agriculture, and medical research. It has applications ranging from creating genetically modified organisms (GMOs) to studying gene function and developing therapeutic treatments.

13.3.1.1. Cry 1

Bacillus thuringiensis (*Bt*) is a type of Gram-positive soil bacterium that produces insecticidal crystalline proteins (ICPs) with particularly potent effects against specific classes of pests (Panwar *et al.*, 2018). In *Bt* crops engineered for insect resistance, the genes responsible for producing parasporal crystal protoxins drive the expression of these insecticidal properties (Palma *et al.*, 2014). The incorporation of ICPs into transgenic plants has significantly impacted the development of insect resistance over time. The crystalline structure contains protoxin proteins that become soluble within the alkaline environment of the larval midgut. Following this, enzymatic cleavage transforms them into active toxins. These toxins then permeate the peritrophic membrane encompassing the gut and interact with receptors found in the midgut epithelium, leading to the creation of pores in the midgut lining (Paul and Das, 2020). As a result, the gut of pest becomes immobilized, leading to the cessation of feeding, and eventual death occurs within a span of 2 to 3 days. The mechanism of *Bt* cry toxin in pest is presented in Fig.2

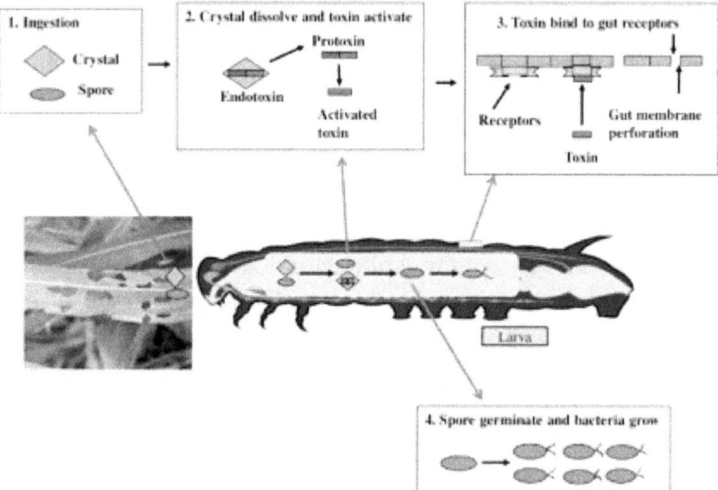

Fig.13.2. Mechanism of Bt Cry Toxin Action in Insects

13.3.1.2. Lectins

Lectins, a group of proteins found in various organisms, have emerged as potential agents for enhancing insect pest resistance in plants. These proteins possess the unique ability to bind to specific carbohydrate molecules, and this characteristic has been harnessed for their application in pest management. When ingested by insects, lectins can exert toxic effects by disrupting insect gut function, interfering with nutrient absorption, and even causing gut paralysis. Some lectins possess insecticidal properties that disrupt vital physiological processes in pests, such as cell membrane integrity and enzyme activity. Researchers have explored the possibility of introducing genes encoding insecticidal lectins into plants through genetic modification. This strategy aims to enable crops to produce lectins within their tissues, deterring pests upon ingestion. This approach offers the potential to reduce reliance on chemical pesticides, promoting more sustainable and environmentally friendly pest control practices. However, challenges such as potential effects on non-target organisms and variability in effectiveness against different pest species warrant careful consideration in the development and deployment of lectin-based insect pest resistance strategies.

Genetically modified rice containing *Allium sativum* leaf agglutinin (ASAL) and *Galanthus nivalis* lectin (GNA) exhibited protection against significant sap-sucking insects such as brown planthoppers (BPH), white-backed planthoppers (WBPH), and green leafhoppers (GLH), as reported by Bharathi *et al.* in 2011. Increased toxicity of the fusion protein GNA-spider venom toxin I (SFI1) towards larvae of the tomato moth (*L. oleracea*), rice brown planthopper (*N. lugens*), and the peach potato aphid (*M. persicae*) was documented in studies conducted by Fitches *et al.* in 2004 and Down *et al.* in 2006.

13.3.1.3 Protease Inhibitors

Protease inhibitors have emerged as a significant component in the realm of insect pest resistance in plants. These molecules play a crucial role in preventing the feeding and digestion processes of pests. Protease inhibitors, as the name suggests, hinder the activity of proteolytic enzymes in the digestive systems of insects. By inhibiting these enzymes, plants can render ingested nutrients less accessible, leading to reduced growth and survival of the pests. The utilization of protease inhibitors as a defense mechanism showcases the intricate strategies plants employ to counter insect attacks. Researchers have explored the incorporation of genes encoding protease inhibitors into crops, thereby equipping them with enhanced resistance against a spectrum of pest species. This approach offers the advantage of reduced dependence on conventional chemical pesticides, promoting more sustainable agricultural practices. However, the effectiveness of protease inhibitors can vary across different pests, and potential interactions with non-target organisms necessitate careful evaluation when applying this approach in pest management strategies.

13.3.1.4. Insect Chitinase

Insect chitinase has gained prominence as a pivotal element in the context of insect pest resistance within plants. Chitinase is an enzyme that targets chitin, a major component of insect exoskeletons and other structural elements. By degrading chitin, insect chitinase disrupts the integrity and stability of pests' physical structures, thereby impeding their growth and development. Plants have evolved to produce chitinase as part of their defense arsenal, and this enzyme is particularly effective against chitin-containing insects.

Through biotechnological approaches, researchers have explored the incorporation of insect chitinase genes into crops. This genetic enhancement empowers plants with the ability to produce chitinase, leading to increased resistance against a range of insect pests. When pests feed on these chitinase-producing plants, their chitin-rich exoskeletons become vulnerable to enzymatic degradation, compromising their ability to molt and grow. This phenomenon results in stunted pest development and reduced survival rates. The incorporation of insect chitinase genes in crops offers a sustainable alternative to traditional chemical pesticides, aligning with the principles of integrated pest management. However, the effectiveness of this approach can vary depending on the targeted pest species, necessitating a comprehensive understanding of the specific pests' susceptibility to chitinase. Furthermore, potential ecological impacts and interactions with non-target organisms must be carefully evaluated to ensure the responsible deployment of insect chitinase for insect pest resistance.

13.4. Marker Assisted Selection

The development of DNA (or molecular) markers has irreversibly changed the disciplines of plant genetics and plant breeding. While there are several applications of DNA markers in breeding, the most promising for cultivar development is called marker assisted selection (MAS). MAS refers to the use of DNA markers that are tightly-linked to target loci as a substitute for or to assist phenotypic screening. By determining the allele of a DNA marker, plants that possess particular genes or quantitative trait loci (QTLs) may be identified based on their genotype rather than their phenotype.

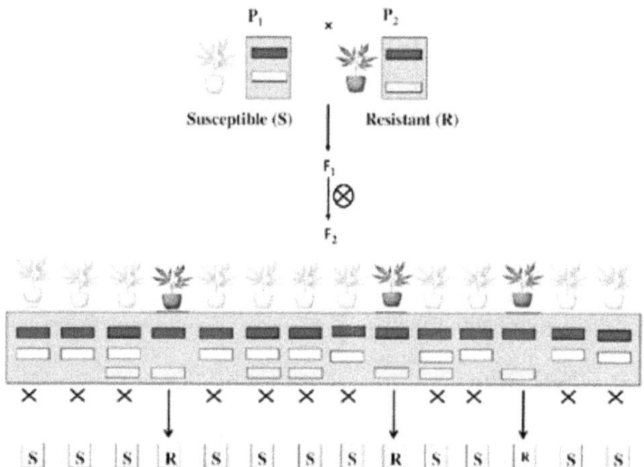

Fig.13.3. Marker Assisted Selection in Insect Pest Management

Marker-assisted selection (MAS) stands as a potent strategy in the pursuit of insect pest-resistant crops. Through this approach, specific genes associated with insect resistance are identified, and molecular markers linked to these genes are pinpointed. By genotyping potential plant candidates, breeders can swiftly and accurately identify those carrying the desired molecular markers. This expedites the selection process, rendering traditional phenotypic assessments less necessary. The markers serve as reliable indicators of pest resistance, enabling breeders to make informed choices while optimizing time and resources. MAS not only accelerates the development of pest-resistant varieties but also facilitates the combination of multiple resistance traits into a single plant. In essence, MAS revolutionizes insect pest resistance breeding by harnessing genetic information to expedite the creation of resilient and productive crop varieties. Application of marker-assisted selection in insect pest management is presented in fig.13.3.

The utilization of molecular marker-assisted selection (MAS) in studies revealed the successful integration of *Bph14* and *Bph15*, enhancing resistance against brown planthopper (BPH) in Minghui 63 and its derived hybrids, as demonstrated by Hu *et al.* in 2012. The rice line ASD7 carrying BPH resistance gene *bph2* when subjected to marker-assisted selection (MAS) and subsequently crossed with the susceptible cultivar C418, a *japonica* restorer line. The resulting progeny, as evaluated by Li-Hong *et al.* in 2006, demonstrated significantly high resistance against the brown planthopper *Nilaparvata lugens* which is an immensely damaging pest to rice crops. Marker-assisted pyramiding was employed by Sharma *et al.*, 2004 to effectively combine the *Bph1* and *Bph2* resistance genes onto rice chromosome 12, resulting in the development of resistance against the rice brown planthopper (BPH).

Utilizing MAS to enhance pest resistance presents significant advantages. Several benefits of employing MAS to improve the efficiency of selecting insect-resistant plants include the ability to apply it to seedling material, its reduced susceptibility to environmental influences, its potential for cost-effective and quicker outcomes compared to traditional phenotypic assessments, and the capacity to evaluate multiple markers using the same DNA sample. However, potential drawbacks should also be noted. Firstly, recombination between the marker and the target gene could lead to inaccurate outcomes. Secondly, incorrect estimations of QTL locations and effects might result in slower advancements than anticipated. Lastly, the transferability of markers developed for MAS from one population to others might be limited.

13.5. Gene editing Techniques

Gene editing techniques have revolutionized insect pest management by providing precise and targeted methods to modify the genetic makeup of insects. These techniques offer innovative approaches for controlling pest populations, minimizing environmental impact, and reducing reliance on chemical pesticides.

13.5.1. CRISPR-Cas9 Technology

CRISPR-Cas9 technology for breeding insect pest resistance in crops involves precise genetic modification to enhance the plant's defenses against specific insect pests. CRISPR-Cas9 is a revolutionary genome editing tool that allows researchers to modify DNA with remarkable precision, enabling the creation of genetically modified organisms (GMOs) with desired traits.

CRISPR-Cas9 has the ability to precisely edit specific genes within an organism's genome. For insect pest resistance, researchers identify key genes responsible for traits such as pesticide resistance, reproduction, or feeding behavior, and then use CRISPR-Cas9 to introduce deliberate alterations. By disrupting or modifying these genes, insects can exhibit reduced fitness, compromise reproductive abilities, or alter behaviors, all of which contribute to limiting their impact on agricultural productivity. CRISPR-Cas9 technology can be applied for breeding insect pest resistance using genomic tools as follows and presented in Fig.4.

- ❒ **Identifying Target Genes:** Identify specific genes in the crop plant that are involved in the defense against the targeted insect pests. These genes might encode proteins responsible for producing toxic compounds or triggering immune responses against the pests.

- ❒ **Designing Guide RNAs:** Design guide RNAs (gRNAs) that target the identified genes. The gRNAs serve as molecular "scissors" that guide the Cas9 enzyme to the precise location within the plant's DNA where the desired gene modification needs to occur.

- ❒ **Gene Editing:** Introduce the CRISPR-Cas9 system into the plant cells. This can be done through various methods such as Agrobacterium-mediated transformation or direct delivery of CRISPR components into the cells. The Cas9 enzyme, guided by the gRNA, then cuts the target gene at the specified location.

- ❒ **Repair Mechanisms:** When the target gene is cut, the plant's cellular repair mechanisms come into play. These mechanisms can result in one of two outcomes: Non-Homologous End Joining (NHEJ) or Homology-Directed Repair (HDR).

- ❒ **NHEJ:** Often leads to small insertions or deletions (indels) at the site of the cut, which can disrupt the functionality of the targeted gene.

- ❒ **HDR:** Involves introducing a DNA repair template that carries the desired genetic modification. This can lead to precise changes in the gene sequence, such as introducing resistance-conferring mutations or altering gene expression.

- ❒ **Screening and Selection:** Screen and select the edited plants for the desired trait. In this case, plants that show enhanced resistance to the targeted insect pests would be identified and further propagated.

- ❒ **Validation and Testing:** Conduct comprehensive testing to ensure that the edited plants are safe for consumption, environmentally sustainable, and that the modified traits are stable across generations.

13.5.1.1 Applications of CRISPR/Cas9 in Insect Pest Management

Kandul *et al.*, 2019 introduced a groundbreaking method known as the precision-guided sterile insect technique (pgSIT), which harnessed CRISPR/Cas9-mediated genome editing in *Drosophila*. The sterile insect technique (SIT) stands as a well-established and environmentally sound approach to curbing insect populations. Multiple pgSIT systems were engineered efciently in *Drosophila* to constantly produce 100% sterile male population. This meticulously designed pgSIT model population demonstrated both competitiveness and remarkable efficacy in suppressing disease vectors and insect pests, thereby underscoring its significant potential. Another avenue of pest management strategy, driven by CRISPR/Cas9 technology was studied by Gui *et al.*, 2020. This research focused on investigating the biology of the Colorado potato beetle (CPB), *Leptinotarsa decemlineata*, a significant pest affecting the *Solanaceae* family, which includes potatoes. The study utilized CRISPR/Cas9-induced mutagenesis on the vestigial gene (*vest*), resulting in the emergence of wingless adult CPBs devoid of elytron formation which was shown in **Fig.13.4.**

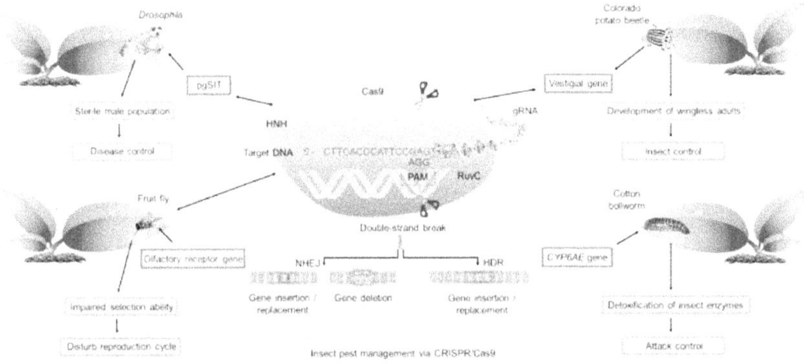

Fig.13.4. CRISPR/Cas9-mediated genome editing applications for insect pest resistance in plants

Furthermore, CRISPR-Cas9 offers a precise method for conferring insecticide resistance to beneficial organisms or non-target species. By introducing specific genetic alterations, these organisms can withstand chemical treatments while pests remain susceptible, minimizing the development of resistance in pest populations. This strategy not only preserves the effectiveness of insecticides but also promotes a balanced ecosystem by safeguarding the survival of beneficial insects that contribute to natural pest control.

While CRISPR-Cas9 presents a promising frontier for insect pest resistance, its implementation is not without challenges. Ethical considerations, potential off-target effects, and the potential for altered ecosystems due to gene modifications warrant careful examination. Regulatory frameworks must be established to ensure the responsible use of this technology, considering both the benefits and potential risks associated with the genetic modification of insect populations.

CRISPR-Cas9 represents a groundbreaking advancement in the quest for effective insect pest resistance. Its precision and versatility hold the potential to reshape pest management strategies, reducing the reliance on conventional pesticides while promoting sustainable and targeted solutions for safeguarding global agriculture. As researchers continue to refine and explore the potential applications of CRISPR-Cas9, it is crucial to strike a balance between innovation and responsible use to address the challenges posed by insect pests in an ever-changing world.

13.6. RNA Interference for Insect Pest Resistance

RNA interference (RNAi) has emerged as a promising and versatile tool in the realm of insect pest resistance, offering a precise and environmentally friendly approach to curbing pest populations. This molecular mechanism exploits the natural process of gene regulation within organisms to control the expression of specific genes, effectively silencing their activity. In the context of insect pest resistance, RNAi allows researchers to target crucial genes essential for pest survival, growth, or reproduction, resulting in reduced pest populations and crop damage.

13.6.1. Mechanism of RNAi

The mechanism of RNAi involves introducing double-stranded RNA (dsRNA) molecules that correspond to specific target genes into the organism. Once inside the pest, these dsRNA molecules trigger a cellular response, leading to the degradation of the corresponding messenger RNA (mRNA) and, consequently, the inhibition of protein synthesis. By silencing genes crucial for pest survival, RNAi can hinder essential biological processes, resulting in decreased pest fitness and, in some cases, mortality.

One notable application of RNAi in insect pest management is the development of RNAi-based insecticides. These insecticides harness the power of RNAi to selectively target pests without harming beneficial organisms. By designing dsRNA molecules that target specific genes unique to pests, these insecticides effectively silence essential genes in the pests, ultimately leading to their demise. Unlike conventional chemical insecticides, RNAi-based solutions are highly specific, minimizing collateral damage to non-target species and reducing environmental impact.

Additionally, RNAi can be exploited to address pesticide resistance issues. By targeting genes associated with insecticide resistance, researchers can potentially reverse or mitigate resistance development in pest populations. This approach not only extends the efficacy of existing insecticides but also offers a sustainable alternative to the continuous use of chemical compounds. Despite its potential, RNAi-based pest resistance is not without challenges. Efficient delivery of dsRNA molecules to the target pests, potential off-target effects, and the persistence of the RNAi effect over time are important factors to consider. Additionally, regulatory considerations surrounding the use of genetically modified organisms and gene-silencing technologies must be addressed for responsible deployment.

13.6.2. Benefits of Using Genomic Tools for Pest and Insect Resistance Breeding Include

- **Precision:** Genomic tools allow for targeted modifications, reducing the likelihood of unintended effects.

- **Efficiency:** Traditional breeding can be time-consuming; genomic tools expedite the process by focusing on specific genes.

- **Reduced Environmental Impact:** Pest-resistant crops can reduce the need for chemical pesticides, leading to decreased environmental pollution.

- **Resilience:** Multiple resistance genes provide a backup in case pests evolve to overcome a single resistance trait.

13.7. Challenges and Future Prospects of Insect Pest Resistance Using Genomic Tools

Breeding for insect pest resistance using genomic tools presents several challenges that must be navigated to ensure successful implementation. One of the primary challenges is the genetic complexity of resistance traits. Insect resistance often involves multiple genes, making it intricate to identify and manipulate the complete set of genetic factors responsible for the desired trait. Additionally, insects have a remarkable ability to adapt and evolve, potentially rendering resistance traits ineffective over time. The rapid adaptation of insect pests can undermine the long-term efficacy of resistant crops, necessitating continuous vigilance and adaptation of breeding strategies. Manipulating the genome to introduce resistance traits might inadvertently affect other essential plant attributes, leading to unforeseen negative impacts on crop yield, quality, or even environmental interactions. Furthermore, the regulatory approval process for genetically modified organisms (GMOs) can be a substantial hurdle. Meeting the stringent requirements and addressing safety concerns associated with genetically modified crops demands significant time and resources. Ethical and environmental concerns also cast a shadow over the development of insect-resistant crops. The potential for genetic modifications to disrupt ecosystems or harm non-target species raises questions about the broader ecological consequences of such interventions. Striking a balance between addressing food security needs and maintaining ecological stability is a complex challenge.

Despite the challenges, breeding for insect pest resistance through genomic tools offers a range of exciting opportunities that can revolutionize agricultural practices. The precision afforded by genomic tools enables the targeted manipulation of specific genes associated with insect resistance. This level of precision minimizes unintended effects on other traits, facilitating the development of crops with enhanced resistance while preserving desired plant characteristics. Traditional breeding methods can be time-intensive, requiring years of crossbreeding and selection. Genomic techniques, however, expedite the process by identifying plants with desired traits at the molecular level. This acceleration is particularly crucial in addressing the urgent need for resilient and sustainable agriculture in the face of changing climate conditions and evolving pest pressures.

Genomic tools also unlock the potential for identifying and characterizing resistance genes through techniques like genome-wide association studies (GWAS) and quantitative trait locus (QTL) mapping. These tools allow breeders to understand the genetic basis of resistance, aiding in the targeted selection and stacking of resistance traits within plant varieties. Moreover, gene editing technologies such as CRISPR-Cas9 offer the ability to introduce or enhance resistance traits without incorporating foreign DNA, addressing some of the regulatory concerns surrounding GMOs. Breeding for insect pest resistance not only reduces reliance on chemical pesticides but also holds the promise of improving crop yield and quality. By mitigating the damage caused by insect pests, resistant crops can achieve higher yields and better overall crop health. This aligns with the growing demand for sustainable and environmentally friendly agricultural practices that minimize chemical inputs.

References

Bharathi, Y., Vijaya Kumar, S., Pasalu, I. C., Balachandran, S. M., Reddy, V. D and Rao, K. V. 2011. Pyramided rice lines harbouring *Allium Sativum* (Asal) and *Galanthus Nivalis* (Gna) lectin genes impart enhanced resistance against major sap sucking pests. *Journal of Biotechnology.* 152 (3): 63–71.

Down, R. E., Fitches, E. C., Wiles, D. P., Corti, P., Bell, H. A., Gatehouse, J. A and John, P.E. 2006. Insecticidal spider venom toxin fused to snowdrop lectin is toxic to the peach-potato aphid, *Myzus persicae* (Hemiptera: Aphididae) and the rice brown planthopper, *Nilaparvata lugens* (Hemiptera: Delphacidae). *Pest Management Science.* 62 (1): 77–85.

El-Dessouki, W. A., Mansour, M. R. K and Eryan, N. L. 2022. Effects of certain weather, biotic factors and chemical components on the population of aphids in Egyptian wheat fields. *Egyptian Academic Journal of Biological Sciences A, Entomology.* 15 (1): 1–13.

Fitches, E., Wilkinson, H., Bell, H., Bown, D. P., Gatehouse, J. A and Edwards, J. P. 2004. Cloning, expression and functional characterisation of chitinase from larvae of tomato moth (*Lacanobia oleracea*): a demonstration of the insecticidal activity of insect chitinase. *Insect Biochemistry and Molecular Biology.* 34 (10): 037–1050.

Gui, S., Taning, C.N.T., Wei, D and Smagghe, G. 2020. First report on CRISPR/ Cas9-targeted mutagenesis in the colorado potato beetle. *Leptinotarsa decemlineata.* Journal of Insect Physiology. 121:10401.

Horgan, F. G., Garcia, C. P. F., Haverkort, F., de Jong, P. W and Ferrater, J. B.2020. Changes in insecticide resistance and host range performance of planthoppers artificially selected to feed on resistant rice. *Crop Protection.* 127.

Hu, J., Li, X., Wu, C., Yang, C., Hua, H., Gao, G., Jinghua, X and Yuqing, H. 2012. Pyramiding and evaluation of the brown planthopper resistance genes *Bph14* and *Bph15* in hybrid rice. *Molecular Breeding.* 29 (1): 61–69.

Jaiswal, D. K., Raju, S. V. S., Kumar, G. S., Sharma, K. R., Singh, D. K and Vennela,P. R. 2018. Biotechnology in plant resistance to insects. *Indian Journal of Agriculture and Allied Sciences.* 4: 7–18.

Joshi, M. J., Prithiv Raj, V and Solanki, C. B. 2020. Role of biotechnology in Insect-pests management. *Agriculture and Food*: e-Newsletter. 2: 574–576.

Kandul N.P., Liu, J, Wu, S.L., Marshall, J.M., Akbari, O.S. 2019. Transforming insect population control with precision guided sterile males with demonstration in flies. *Nature Communications.* 10:1–12.

Kumari, P., Jasrotia, P., Kumar, D., Kashyap, P.L., Kumar, S., Mishra, C.N., Kumar, S and Singh, G.P. 2022. Biotechnological approaches for host plant resistance to insect pests. *Frontiers in Genetics.* 13:914029.

Li-Hong, S. U. N., Chun-Ming, W. A. N. G., Chang-Chao, S. U., Yu-Qiang, L. I. U., Hu-Qu, Z. H. A. I., and Jian-Min, W. A. N. 2006. Mapping and marker-assisted selection of a brown planthopper resistance gene bph2 in rice (*Oryza sativa* L.). *Acta Genetica Sinica.* 33 (8): 717–723.

Palma, L., Munoz, D., Berry, C., Murillo, J and Caballero, P. 2014. *Bacillus thuringiensis* toxins: an overview of their biocidal activity. *Toxins.*6 (12): 3296–3325.

Panwar, B. S., Ram, C., Narula, R. K and Kaur, S. 2018. Pool deconvolution approach for high-throughput gene mining from *Bacillus thuringiensis.* *International Journal of Applied Microbiology and Biotechnology Research.*102 (3): 1467–1482.

Paul, S and Das, S. 2020. Natural insecticidal proteins, the promising biocontrol compounds for future crop protection. *Nucleus.* 64 (1): 7–20.

Sharma, P. N., Torii, A., Takumi, S., Mori, N and Nakamura, C. 2004. Marker-assisted pyramiding of brown planthopper (*Nilaparvata Lugens*) resistance genes Bph1 and Bph2 on Rice Chromosome 12. *Hereditas.* 140 (1): 61–69.

Chapter-14

Improving Health Benefits Through Plant Breeding: Breeding For Nutritional Quality

R. Nivedha[1] and T. Nivethitha[1]

Department of Genetics and Plant Breeding, Tamil Nadu Agricultural University

Corresponding author: nivedharakkimuthu@gmail.com

14.1. Introduction

The main area of plant breeding is concentrated on the increase in yield to meet the demand of ever-growing population. On the other hand, achieving self-sufficiency alone does not meet the very purpose unless it is not compromised for quality. A well-balanced diet is of paramount value in determining good health and development of humans. According to World Health Organization, India holds the third position in undernutrition. Hidden hunger known as micronutrient deficiency resulting from energy dense but nutrient poor diet causes mental impairment, poor health and productivity [7]. This necessitates the focus on the nutrition aspects of crops which are the main source of carbohydrates, fats, proteins, vitamins and other minerals. Biofortification is the cost-effective method of improving the nutritional quality in crops through conventional breeding or biotechnological approaches. This can address millions of people who cannot afford nutritionally diverse diet or chemical supplements. HarvestPlus is a part of Consultative Group for International Agricultural Research which focuses on sustainable biofortification of staple crops to improve the nutrition status of under developed sector. The following chapter focuses on the breeding methods and nutritional quality improvement of major food crops.

14.2. Genetics of Nutritional Traits

The nutritional traits are governed by oligogenes, polygenes and a few by maternal effects. The oligogenic traits controlled by one or few genes are lysine content in cereals, amylose in rice, carotene in tomato and trypsin in soybean. The traits showing polygenic inheritance with additive gene action are fibre length and strength in cotton, protein content in pulses, carotene content in carrot and grain starch in wheat. Maternal effects are generally known in seed characters and for protein content in chickpea. They reduce the progress of selection because of the confusion between genotypic and phenotypic effect [11].

14.3. Biofortification Approaches

- ☐ Agronomic practices
- ☐ Conventional breeding
- ☐ Biotechnological approach

14.3.1. Agronomic Practices

Apart from macronutrients, plants require various micronutrients like iron, zinc, iodine, copper, manganese, molybdenum *etc*. The agronomic biofortification involves soil and foliar application of micronutrients which can be easily absorbed by plants. In some cases, where plants are unable to absorb, microbes like rhizobium, azotobacter, bacillus and actinomycetes are utilized to improve nutrient availability to plants.

14.3.2. Conventional Breeding

The conventional breeding requires sufficient genetic variation for the trait of interest. From the germplasm, lines with high nutrient content were identified and crossed with recipient parent with desirable agronomic performance to obtain high yielding nutritionally enriched hybrids. These are further selfed and forwarded to several generations to follow recurrent selection in the segregating generations in case of polygenic traits. If sufficient variation is not present for the desirable nutrient trait, interspecific hybridization by crossing with distant parent can be useful. In addition to this, mutagenesis can be employed to introduce new trait to the existing varieties.

14.3.3. Biotechnological Approach

The application of molecular markers to identify QTLs governing nutritional traits and efficient transfer of polygenic traits through marker assisted selection is more advantageous. When there is no variation in the existing gene pool for the nutrient trait, biotechnological methods like somaclonal variation, transgenic breeding, genome editing and omics approach are used in biofortification of crops. The micronutrients and vitamins can be increased by transferring genes from other genomes into plants or by overexpression of the genes responsible for their synthesis. Example, the overexpression of *TaFer1-A* and *OsNAS2* genes increased iron content upto 80-85 µg/g in wheat grain [3], Golden rice has genes introduced from daffodil and bacterium to enable biosynthesis of provitamin A. The presence of anti-nutritional factors in pulses poses nutritional hinderance which can be rectified by suppression of gene expression by RNA interference technology. Genome editing techniques like Crispr-cas9, TALENS were employed to alter a particular region of the genome. The iron availability was enhanced in rice by knocking down the gene *OsVIT2* through genome editing (complete book). The omics approaches like transcriptomics, proteomics and metabolomics involve measuring biological molecules like RNA, protein and analyzing the expression level of transgenics to reveal the molecular mechanisms to enhance the nutrional quality of crops.

14.4. Breeding for Nutrition in Cereals

14.4.1. Rice

14.4.1.1. Pigmented Varieties

Being the most prominent staple crop, globally, rice contributes about 25% of energy requirement [5]. Brown rice is more nutritious than white rice because of the presence of vitamins, minerals, antioxidants and fatty acids in the outer bran layer. The pigmented black and purple rice contains bioactive compounds like anthocyanin, flavonoids and phenolic acids with health benefits. The traditional pigmented varieties can be used as one of the parents in crossing to obtain nutritionally rich varieties. In Indonesia, high yielding red rice variety Pamelen was produced by an interspecific cross between *Oryza rufipogon* and IR64 [4]. The Tamil Nadu Agricultural University has released an improved variety of kavuni (black rice) called CO 57. It has high fibre (3-3.5%), low carbohydrates (65-70%) and medium glycemic index (67%) than other varieties [18].

14.4.1.2. Protein

In rice, 20% of protein is found in embryo and aleurine layer and 80% is contained in endosperm. The average amount of protein found in polished rice is 7%. The rice protein is considered incomplete because of the deficiency in lysine. By modifying the pathway of lysin *i.e.,* by depressing the lysine catabolism, *Lys* is reported to be increased in rice [14]. In the other way, over expression of lysine rich proteins also helped in increasing the *Lys* content in rice. The rice varieties with high protein are CR Dhan 310 (10.3%), CR Dhan 411 (high protein Swarna with 10% protein) and CR Dhan 311 (10.1%).

14.4.1.3. Provitamin A

Rice cultivars lack biosynthetic pathway to produce provitamin A which are converted to vitamin A in humans. Wild type rice can synthesize geranylgeranyl diphosphate (GGPP) which is the precursor of carotenoids and lack other enzymes *viz.*, phytoene synthase (PSY), phytoene desaturase (PDS), ζ-carotene desaturase (ZDS), carotene cis–trans-isomerase (CRTISO) and lycopene-β cyclase (LCYB) to produce β carotene from GGPP. In 2000, Ingo Potrykus and Peter bayer developed genetically engineered golden rice which produce β carotene, precursor of vitamin A in the grain endosperm. The production of carotenoids gives the yellow colour to the golden rice from which it gets its name (*Figure 1*). In golden rice 1 (GR1), *psy* gene from daffodil encoding phytoene synthase and *crtI* gene from the bacterium *Pantoea anamatis* (previously called *Erwinia uredovora*) encoding phytoene desaturase were genetically introduced into rice cultivar. This accounted for 1.6µg of carotenoids per gram of grain. Later, the rate limiting psy gene from daffodil was replaced by its homolog from maize that comparatively increased carotenoid accumulation to 23-fold in golden rice 2 (GR2) which accounted for 37 µg/g of carotenoid [17]. The later version of GR contains tHMG1 gene from *Saccharomyces cerevisiae* in addition to psy and crtI which was reported to have 63 fold increase in β carotene in the endosperm [5].

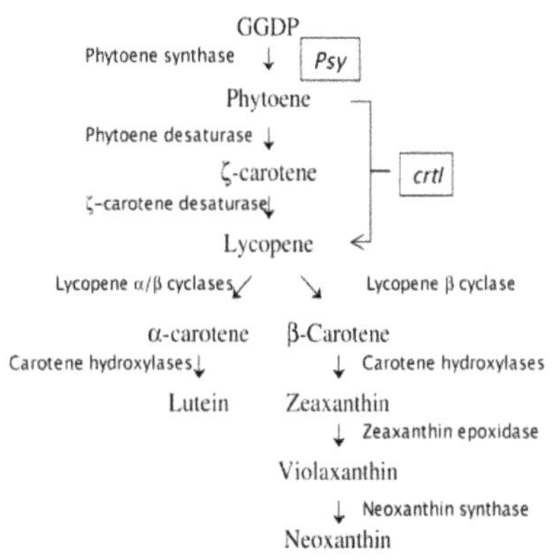

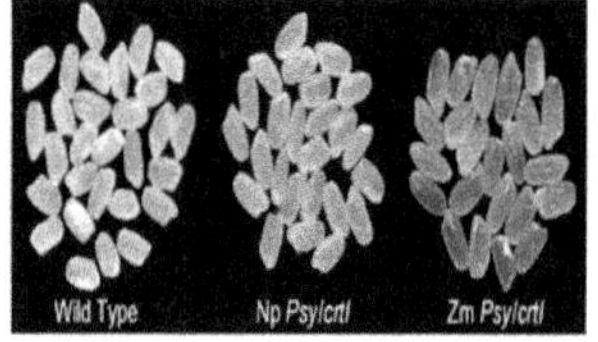

The pathway of β carotene synthesis in transgenic rice [17]

14.4.1.4. Micronutrients

The unequal distribution of micronutrients in plants with minimal nutrient in tissues of edible parts rather than leaves accounted for poor nutrient. In rice, leaves contain all the necessary micronutrients while the grains have minimal quantity [15]. The micronutrient zinc is accumulated in the endosperm whereas iron is accumulated in the aleurone layer which gets lost in the range of 16-97.4% during milling. The average amount of iron and zinc content in polished rice is approximately 2 µg and 16 µg respectively. The rice germplasm has wide variations for iron and zinc content thus enabling their improvement through conventional breeding. The high Zn rice varieties are CR Dhan 315 (25 ppm), DRR Dhan 48 (22 ppm) and DRR Dhan 49 (25.2ppm).

14.4.2. Wheat

14.4.2.1. Pigmented varieties

The coloured wheat exists in three colours *viz.*, purple, blue and black based on the presence of anthocyanin in pericarp layer, aleurone layer and mixture of pericarp and aleurone layer respectively. The coloured wheat has the advantage of bearing higher concentration of micronutrients and high stability of aminoacids upon cooking than common wheat [6]. Few anthocyanin rich wheat varieties namely NABIMG9, NABIMG 10 and NABIMG 11 having improved Zn content were developed.

14.4.2.2. Protein and micronutrients

The improvement in protein and micronutrient contents is the major area of focus in wheat biofortification. Protein is a quantitative trait affected by environmental factors which is negatively correlated with yield. The average protein content in wheat is 12.4% (BD Singh). Atlas 66 is a source of high protein without affecting the yield. It was used to develop the variety Lancota with 1-2% more protein. In case of micronutrients, the wheat germplasm has wide diversity for iron and zinc content. Zinc Shakti is a high Zn (14ppm) wheat variety developed by participatory variety selection. The ICAR developed a high iron wheat variety HD3171 with the iron content of 47.2 ppm [8].

14.4.3. Maize

14.4.3.1. Protein

The maize protein is deficient in essential amino acids lysine and tryptophan. Among 8-10% protein, the lysin content is 1.5-2%. But the daily requirement of lysine is 30mg/kg and 35mg/kg for adult and children respectively. The daily requirement of tryptophan for adult and children is 4mg/kg and 4.8mg/kg respectively. Maize protein has high leucine isoleucine ratio and 60% of protein fraction is composed of zein proteins (deficient in lysine and tryptophan) and the remaining is occupied by non-zein proteins. The improvement in the nutritional quality of maize protein lead to the development of Quality Protein Maize (QPM). Many spontaneous natural mutations have been reported in maize kernel *viz.*, *opaque2 (o2)*, *opaque6 (o6)*, *opaque7 (o7)*, *opaque11 (o11)*, *opaque16 (o16)*, *floury2 (fl2)*, *floury3 (fl3)*, *Mucronate (Mc) and Defective endosperm (De-B30)* [12]. The o2 mutant causes reduction in the expression of zein proteins and enhancement in the synthesis of lysin rich non-zein proteins. Also, the transcription of the enzyme lysine keto reductase which degrades maize lysine is reduced. This accounted for about two-fold increase in lysin and tryptophan than traditional maize. The reduction in zein proteins causes reduction in the protein bodies which results in loose packaging leading to soft and opaque kernels. In addition to this, o2 mutant leads to low kernel density, soft kernel impeding mechanical processing. To overcome this, double mutant combinations *viz.*, o2/fl2, o2/

su2 were tried but was not successful. The endosperm modifier genes which cause the soft kernels to get converted into vitreous kernels were identified and accumulated in o2 genetic background leading to the development of Quality Protein Maize. QPM has many advantages like high biological value (80%), decreased leucine isoleucine ratio and more tryptophan for niacin synthesis. The biofortified o2 based maize composites released in India are Shakti, Rattan and Protina . Endosperm modifier based QPM composite is Shakti-1. The first QPM hybrid developed through three-way cross was Shaktiman-1. The first marker assisted backcross breeding was carried out to transfer o2 allele from QPM inbred with white kernels CML176 into V25 an yellow kernel inbred. The first cultivar bred through Marker Assisted Selection released for commercial cultivation in India is Vivek QPM 9. It was developed by improving the lysin and trptophan content of the parents of Vivek hybrid-9. It had 41% more tryptophan and 30% more lysine than the original hybrid. The other cultivar developed by MAS is Vivek QPM 21[12]. The mutant combination of o16 and o2 resulted in higher lysine concentration than the presence of o2 mutant alone. So, commercial introgression of o16 into QPM hybrids offers better way of nutrition enhancement in maize.

14.4.3.2. Provitamin A

The low β carotene in maize germplasm invited the use of donors containing *lcyE* and *crtRB1* alleles from CIMMYT and HarvestPlus genotypes. Few inbreds which are the parents of commercial maize hybrids (Vivek QPM 9, Vivek Hybrid 27, HM4 and HM8) were introgressed with *crtRB1* allele through Marker Assisted Selection. The hybrids were reconstituted from those improved inbreds and were observed to posses 21.7 μg/g of β carotene compared to 2.6 μg/g in original hybrid [13]. The improved Vivek QPM 9 is enriched with multinutrients *viz.*, lysine, tryptophan and β carotene. Similarly, the hybrids APQH-10 and APQH-11 were reconstituted from the improved inbreds incorporated with the favourable alleles of *vte4*, *crtRB1* and *opaque2* genes using molecular breeding. The new hybrids were enriched with β carotene, β cryptoxanthin, lysine and tryptophan [2]. The transgenic approach involved introgression of *crtB* (phytoene synthase) and *crtI* (carotene desaturase) from *Erwinia herbicola* into maize causing an accumulation of 10μg/g of β carotene. Transgenic plants having *psy1* from maize and *crtI* from *Pantoea ananatis* produced as high as 60μg/g of β carotene [16].

14.4.3.3. Micronutrient Improvement in Millets

The International Crop Research Institute for Semi Arid Tropics developed an improved sorghum variety Parbhani Shakti (ICSR 14001) through conventional breeding which has 45 ppm of iron and 32 ppm of zinc. In addition to this it has high protein (11.9%) and low phytate (4.14 mg/100 g) than many other cultivars which further increased the nutrient availability upon consumption [19]. The first iron biofortified pearl millet cultivar in India is Dhanshakti with 71 mg/g iron.

14.4.4. Oil Quality

The oil quality is determined by the type of fatty acids in the triglycerides. The fatty acids with no double bond in the carbon skeleton are termed saturated fatty acids and those with 1, 2 or more double bonds are called unsaturated fatty acids. The fatty acids present in seed oils are given in the following table

Unsaturated fatty acids	Saturated fatty acids
Oleic acid (18:1)	Lauric acid (12:0)
Linoleic acid (18:2)	Palmitic acid (16:0)
Linolenic acid (18:3)	Myristic acid (14:0)
Erucic acid (22:1)	Stearic acid (18:0)
Eicosenoic acid (20:1)	

The presence of high oleic and lenolic acids are considered beneficial for lowering blood cholesterol, controlling blood pressure and heart activity. Breeding for increased oil content was achieved by conventional methods. Example, in sunflower, the oil content was increased from 30% to 54% by recurrent selection. The sunflower has 19% oleic acid and 67% linoleic acid. Both are affected by temperature such that, oleic acid is positively correlated and linoleic acid is negatively correlated with temperature. Breeding for high oleic and linoleic is the primary objective in sunflower. Other crops with good oil quality are groundnut, sesame, safflower and mustard.

14.4.4.1. Antinutritional Factors

The bioavailability of nutrients from plant-based food should be given as much importance as the biofortification of crops. The presence of antinutritional factors in plant products interfere with digestion and the nutrient uptake during consumption through enzyme inhibition or chelation. The elimination or reduction of the concentration of antinutritional compounds is crucial in breeding plants for enhancing nutrition. Among food crops, pulses contain comparatively higher concentration of antinutritional factors than other crops. The traditional processes followed in households to reduce these compounds include soaking, sprouting, roasting, fermenting and boiling. Some of the antinutritional factors and the crop plants in which they are present is given in the following table.

Antinutritional compound	Crop
Phytic acid	Cereals, maize
Saponins	Potato (Solanaceae, liliaceae)
Gossypol	Cotton
Erucic acid	brassica
Oxalic acid	Sugar beet, amaranthus
Vicine	Faba bean
Tannins	Apple, banana, grapes, sorghum
Goitrogen and glucosinolates	Cabbage, rapeseed, canola

Antinutritional compound	Crop
Raffinose	Asparagous, broccoli, beans, brussels sprouts, cabbage, monocot seeds
BOAA	Lathyrus sativus
HCN	Sorghum

14.5. Breeding Methods to Reduce Antinutritional Factors

14.5.1. Conventional Methods

The first attempt to reduce antinutritional compound was made by conventional breeding of screening germplasms in cotton for glandlessness to produce plants with reduced gossypol in seeds [1]. The antinutritional compound BOAA present in kesari dhal is a neurotoxin affecting lower limbs and central nervous system. Screening of lathyrus germplasm identified accessions with low BOAA (eg. P-24) which could be used as parents for breeding. Biotechnological approach of somaclonal variation developed a somaclonal variant called 'Ratan' with low BOAA content in lathyrus. In mustard, presence of erucic acid and glucosinolate causes heart ailments and release toxic goitrogenic products thus hindering its use as animal feed. The internationally accepted standard for erucic acid is it should be <2% in seed oil and for glucosinolates it is <30 ppm in defatted seed meal [10]. The pedigree method of breeding was used to develop the first double zero Indian mustard variety Pusa Double Zero Mustard 31 by ICAR in 2017. The cross between Pusa Mustard-21 which is an Indian mustard variety with low erucic acid and the germplasm accession NUDHYJ-3 with double low content of erucic acid and glucosinolate was carried out and the single plants with low erucic acid in oil combined with low glucosinolate in defatted seed meal were selected by pedigree method in the segregating generations. Mutation breeding is a useful method to combat the negative pleotropic effect of reduced antinutritional factors on yield. The identification of *lpa* mutants in maize lead to the development of low phytic acid lines.

14.5.2. Biotechnological Methods

The marker assisted selection of the interspecific cross *B.rapa x B.oleraceae* helped to alter the polygenic trait glucosinolate in *B.oleraceae*. The antinutritional factor Kunitz Trypsin Inhibitor (*KTI*) in soybean was reduced by marker assisted backcross breeding to develop NRC 142 variety devoid of *KTI* [8]. The advanced genetic approach involves silencing of the genes responsible for synthesizing antinutritional factors by RNAi technology. In wheat, RNAi was utilized to knock out the gene inositol pentakisphosphate kinase (*TalPK*) involved in the biosynthesis pathway of phytic acid [3]. The gossypol content in seeds, foliage and floral organs of cotton was reduced by silencing the gene δ-cadinene synthase. In soybean, the galactinol synthase genes *GmGoLS1A* and *GmGoLS1B* were knock down by two guide RNAs to reduce the concentration of raffinose. The content decreased upto 35% from 64.70mg/g to 41.95mg/g [9]. The *KTI* in soybean was reduced particularly from seeds alone by targeting two genes viz., *KTI1* and *KTI3* involving deletions and insertions in the open reading frame of the genes.

14.6. Conclusion

The biofortification of food crops is the cheapest way of improving rural health of developing countries. The conventional breeding strategies enable the utilization of available variability to enhance the nutrient content and the modern biotechnological approaches help in the specific gene-based modifications without yield penalty. Integration of both approaches play an efficient role in combating micronutrient deficiencies to address malnutrition status of the country.

References

Duraiswamy A, Sneha A NM, Jebakani K S, Selvaraj S, Pramitha J L, Selvaraj R, Petchiammal K I, Kather Sheriff S, Thinakaran J, Rathinamoorthy S, Kumar P R. Genetic manipulation of anti-nutritional factors in major crops for a sustainable diet in future. Frontiers in Plant Science. 2023.13:1070398.

Hossain F, Jaiswal SK, Muthusamy V, Zunjare RU, Mishra SJ, Chand G, Bhatt V, Bhat JS, Das AK, Chauhan HS, Gupta HS. Enhancement of nutritional quality in maize kernel through marker-assisted breeding for vte4, crtRB1, and opaque2 genes. Journal of Applied Genetics. 2023,14:1-3.

Singh N A. Modern concepts in Agricultural Sciences, Integrated Publications, New Delhi, 2023, 25-38.

Sitaresmi T, Hairmansis A, Widyastuti Y, Susanto U, Wibowo BP, Widiastuti ML, Rumanti IA, Suwarno WB, Nugraha Y. Advances in the development of rice varieties with better nutritional quality in Indonesia. Journal of Agriculture and Food Research. 2023. 12:100602.

Zafar S, Jianlong XU. Recent Advances to Enhance Nutritional Quality of Rice. Rice Science. 2023. 30(6):4.

Garg M, Kaur S, Sharma A, Kumari A, Tiwari V, Sharma S, Kapoor P, Sheoran B, Goyal A, Krishania M. Rising demand for healthy foods-anthocyanin biofortified colored wheat is a new research trend. Frontiers in Nutrition. 2022. 9:878221.

Lowe N M. The global challenge of hidden hunger: Perspectives from the field. *Proceedings of the Nutrition Society.* 2021. *80*(3), 283-289.

Gaikwad KB, Rani S, Kumar M, Gupta V, Babu PH, Bainsla NK, Yadav R. Enhancing the nutritional quality of major food crops through conventional and genomics-assisted breeding. Frontiers in Nutrition. 2020. 7:533453.

Le H, Nguyen NH, Ta DT, Le TN, Bui TP, Le NT, Nguyen CX, Rolletschek H, Stacey G, Stacey MG, Pham NB. CRISPR/Cas9-mediated knockout of galactinol synthase-encoding genes reduces raffinose family oligosaccharide levels in soybean seeds. Frontiers in plant science. 2020. 11:2033.

Yadava DK, Yashpal, Vasudev S, Singh N, Saini N, Prabhu KV, Yadav MS, Dhillon MK, Giri SC, Dass B, Singh R. Indian mustard Variety Pusa Double Zero Mustard-31 (PDZM-31). Indian Journal of Genetics and Plant Breeding. 2019. 79(3):636-637

Singh B D. Plant Breeding Principles and Methods. Kalyani Publishers, New Delhi, 2016, 547.

Gupta HS, Hossain F, Muthusamy V. Biofortification of maize: An Indian perspective. Indian Journal of Genetics and Plant Breeding. 2015. 75(01):1-22.

Muthusamy V, Hossain F, Thirunavukkarasu N, Choudhary M, Saha S, Bhat JS, Prasanna BM, Gupta HS. Development of β-carotene rich maize hybrids through marker-assisted introgression of β-carotene hydroxylase allele. PLoS One. 2014. 9(12):e113583.

Long X, Liu Q, Chan M, Wang Q, Sun SS. Metabolic engineering and profiling of rice with increased lysine. Plant Biotechnology Journal. 2013. (4):490-501.

Beyer P. Golden Rice and 'Golden'crops for human nutrition. New Biotechnology. 2010. 27(5):478-81.

Naqvi S, Zhu C, Farre G, Ramessar K, Bassie L, Breitenbach J, Perez Conesa D, Ros G, Sandmann G, Capell T, Christou P. Transgenic multivitamin corn through biofortification of endosperm with three vitamins representing three distinct metabolic pathways. Proceedings of the National Academy of Sciences. 2009. 106(19):7762-7.

Paine JA, Shipton CA, Chaggar S, Howells RM, Kennedy MJ, Vernon G, Wright SY, Hinchliffe E, Adams JL, Silverstone AL, Drake R. Improving the nutritional value of Golden Rice through increased pro-vitamin A content. Nature biotechnology. 2005. 23(4):482-7.

https://agritech.tnau.ac.in/agriculture/CO%2057.html

https://www.icrisat.org/india-gets-its-first-biofortified-sorghum/

Introduction: Novel Approaches in Crop Improvement: Harnessing Cytogenetic Tools

V. Preeti Kumari, Selvamani. S, Ragulakollu Sravanthi and Sathish Kumar. R

Ph.D. Scholar, Department of Genetics and Plant Breeding, TNAU, Coimbatore.

15.1 Introduction

Agriculture is undergoing a transformative revolution, driven by the urgent need to feed a global population that is estimated to reach 9 billion by 2050. Crops are necessary for ensuring the world's food security and economic stability. In Modern era was characterized by increasing global population and the persistent challenges of climate change, sustainable agriculture is of paramount importance for crop improvement. The study of chromosomes, cytogenetics, has a long history entwined with agriculture. Cytogenetic tools are the one of the technologies employed, which involves the study and manipulation of chromosomes, offer an alternative way for accomplishing this goal. These tools delve into the intricate world of chromosomes, where the blueprint of plant life is encoded, offering novel and promising avenues to revolutionize crop breeding and enhance agricultural resilience.

By harnessing the power of cytogenetic tools, scientists are gaining unparalleled insights into the genetic architecture of crops, enabling them to precisely manipulate and augment these fundamental building blocks of life.

These tools include genome editing techniques like CRISPR-Cas9, which provide unparalleled precision in gene modification, to the transfer of alien genes from wild relatives, expanding the genetic repertoire of crops, and the nuanced world of chromosomal mapping, chromosomal engineering and also the manipulation of ploidy levels in crops, each offering unique pathways to crop enhancement. These tools empower us to develop crops that can thrive in an ever-changing world, resist emerging pests and diseases, and meet the nutritional demands of a growing population, and also represents.

This chapter explores the novel applications of cytogenetic tools in crop improvement, highlighting their potential to enhance yield, quality, and resilience.

15.2. Fluorescence in Situ Hybridisation

The introduction of in situ hybridization, cytogenetics was stepped in the molecular era. These techniques locate the DNA sequence in the specific site of the chromosome. The fluorescent probes were used to hybridize the DNA sequence known as FISH. Now a days, various protocols were available in FISH which are useful to diagnose the abnormalities found in the chromosomes.

Fluorescence in situ hybridisation is a type of molecular cytogenetic tool which utilize fluorescent labelled probes binding the specific part of the chromosome to display a high degree of sequence complementarity. Fluorescence microscopy helps to identify where fluorescent probe attached to the specific site of the chromosome. This approach paves a novel way to map and visualize the genetic material in an individual cell, and also to locate the precise genes or part of the gene. It is an indispensable technique for knowing the various kind of chromosomal abnormalities and other mutation.

15.2.1. Types

- **Locus Specific Probes:** These probes bind to the specific part of the chromosome. This kind of probes is useful to find the particular location of gene on the chromosome.

- **Alphoid or Centromere Repeat Probes:** These probes are developed from repetitive sequences which is found in the middle part of every chromosome. The researchers utilise these probes for finding the individual's chromosome number whether they have specific number of chromosome or not. Alphoid probes along with locus specific probes determine the particular chromosome's missing segment.

- **Whole Chromosome Probes:** The collections of smaller probes are called whole chromosome probes and each of them attaches to the different sequence in a given chromosome. By mixing the multiple labelled probes and diverse fluorescent dyes, researchers label every chromosome in distinctive color. The outcome of the chromosome's-coloured map is called spectral karyotype. WCP is useful specifically for analysing the chromosomal abnormalities.

15.2.2. Principle

The fundamental process involves hybridizing nuclear DNA with a nucleic acid probe using either interphase cells or metaphase chromosomes fixed to a microscope slide. The probes are either directly labelled by incorporating a fluorophore or indirectly by incorporating a hapten. Following denaturation, the labelled probe and the target DNA are combined, allowing complementary DNA sequences to anneal. The imaging of the nonfluorescent hapten will need to be done in a separate enzymatic or immunological detection step if the probe had been indirectly labelled. The signals are then examined using fluorescence microscopy. Fluorochrome, which generates colourful signals at the hybridization site, is a component of the enzymatic detection system. The immunological detection method relies on antibodies attaching to certain antigens, which is subsequently shown by a coloured histochemical reaction that may be seen under a light microscope or with fluorochromes under UV light. The most often employed reporter molecules for direct detection are rhodamine, Cy2, Cy3, Cy5, Texas Red and AMCA, as well as fluorescein (fluorescein isothiocyanate or FITC). Biotin, dinitrophenol, and digoxigenin are the most common reporter molecules utilized for indirect detection methods.

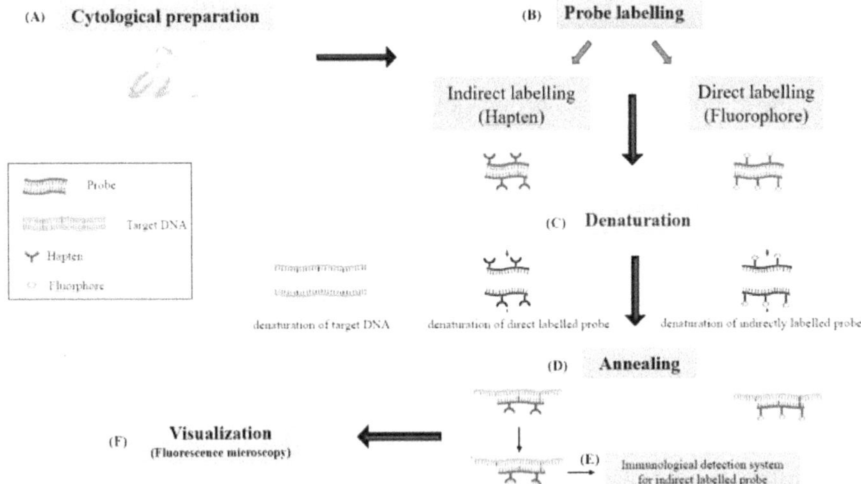

The principle of FISH. The requirements – a target sequence and DNA Probe. (A) Cytological preparation for hybridization (B) First, the DNA probe is indirectly labelled with a hapten or labelled directly with incorporation of fluorophore. (C) The target DNA and labelled probe are denatured to produce a single strand DNA. (D) The annealing process allows the pairing of complementary sequences between target DNA and labelled probe. (E) If the probe has been indirectly labelled, an

15.2.3. Applications

- ❏ FISH can be used as powerful tool to find the location of specific genes
- ❏ It provides diagnostic tool for identifying the chromosomal abnormalities
- ❏ It paves a way to paint the entire chromosomes by utilising the collections of FISH probes
- ❏ It helps to analyse the interphase chromosomes
- ❏ In addition, it has many applications in clinic and research laboratory

15.3. Genomic *In situ* Hybridization (GISH)

Genomic in situ hybridization (GISH) is one of the most important cytogenetic techniques which uses molecular tools for studying plant genomes. Through GISH we can identify homologous and non-homologous chromosomes by hybridizing their genomic DNA. GISH is a kind of modification of basic FISH technique (uses specific DNA sequences as probes) except using total genomic DNA as a probe. The labeling of entire genomes in GISH used for the identification of genomes and in meiotic analysis. The GISH technique involves the isolation of total genomic DNA from the species of interest, which is labeled with a fluorescent dye. The labeled DNA is then hybridized to the chromosomes of the target species, which have been denatured to allow the DNA to bind. The hybridized chromosomes are then visualized under a fluorescence microscope, allowing the identification of homologous and non-homologous chromosomes. GISH been widely used to study origin and evolution of polyploid species, identifying alien chromosomes in hybrids, and for analysing the meiotic behavior of hybrids and polyploids. In addition, GISH also been used in genetic improvement programs, such as pyramiding desirable traits from wild relatives into cultivated crops.

15.4. Chromosome Engineering

Chromosome engineering has emerged as a cutting-edge tool with enormous potential in the pursuit of sustainable agriculture and global food security. This novel method enables researchers and breeders to accurately adjust a plant's genetic makeup by changing its chromosomes, adding or improving desirable features. Chromosome engineering, which involves the translocation and deletion of certain chromosomal segments, provides a means of increasing genetic diversity and developing more robust, productive, and sustainable crops.

15.4.1. Translocation

This procedure allows for the transfer of chromosomal segments between chromosomes. It allows for the introduction of desirable genes or gene combinations from wild relatives or similar species into cultivated crops. For example, if a wild relative has a disease resistance gene, this gene can be translocated onto the chromosome of a cultivated crop and confer the same traits. This technique efficiently increases the crop's genetic diversity, potentially endowing it with resistance to pests, diseases, and abiotic stress.

15.4.2. Deletion

Deletion, on the other hand, is the deliberate removal of certain chromosomal segments. Breeders can increase the yield of crops by removing troublesome regions associated with undesirable or unfavourable characteristics in a crop. This could include increased resistance to diseases and pests, as well as an enhancement of some desirable character.

Chromosome engineering has already had a considerable impact on crops such as wheat, rice, and maize. The major objective of these applications in transmission of traits *viz.*, disease resistance, drought tolerance, and increased nutritional content. As genetic engineering advances, tools such as CRISPR-Cas9 hold the possibility of improving the precision and efficiency of chromosome editing.

In an era defined by climate change and ever-increasing global food requirements, chromosomal engineering emerges as a light of hope. It offers the ability to generate crops that are more robust and also capable of delivering long-term solutions to the complex difficulties that modern agriculture faces. This forward-thinking strategy demonstrates the ability of genetic manipulation in determining the future of food production and maintaining global food security.

Schmidt's groundbreaking research has revealed a previously unexplored aspect of genome engineering. Beyond the precise editing of individual genes, it demonstrates that genome engineering could make the targeted modification of chromosome structure. The core idea is to strategically introduce numerous DNA double-strand breaks (DSBs) into an organism's genome, potentially leading to controlled chromosome modifications. Inducing two DSBs on the same chromosome, for example, could result in deletions or inversions within the region between the breaks, essentially altering the structure of that chromosome. In contrast, intrachromosomal alterations would entail the creation of two or more DSBs on separate chromosomes, potentially resulting in somatic or meiotic crossovers, depending on tissue type and whether breaks occur on one or both homologous chromosomes.

This concept's practical uses are already visible in scientific research. Researchers in Arabidopsis successfully inherited 18 kb inversions by using an egg cell-specific promoter (EC1.1) for Cas9 (SaCas9) expression. Inversions and deletions occurred more frequently in a ku70-1 mutant than in the wild type, indicating improved efficiency via mutagenic microhomology-mediated backup End Joining (EJ). This behaviour was observed in mice as well.

Schmidt's findings have far-reaching ramifications beyond the laboratory. There is an opportunity for reversing natural inversions between closely related agricultural species, facilitating the exchange of beneficial genes. This approach has chance of improving the trait transmission is increasing between species and demonstrates the ever-expanding boundaries of genetic research and biotechnology.

15.5. Alien Gene Introgression

Alien gene introgression, also referred to as interspecific or intergeneric hybridization, is a sophisticated and novel genetic approach used in plant breeding. The transfer of genes or chromosomal segments from wild plant relatives or closely related species into cultivated crops is the primary objective of this strategy. The main objective is to introduce novel genetic material into crop plants, conferring favorable features such as disease resistance, stress tolerance, increased yield, or increased nutritional value. Mechanism and Procedure:

Alien gene introgression unfolds through a meticulously designed and executed multi-step process:

15.5.1. Selection of Donor Species

The journey begins with the careful selection of a donor species or wild relative famous for exhibiting the traits that are wanted. These characteristics may include resistance to specific pests or diseases, the ability to survive environmental challenges such as drought or salinity, or even a proclivity to thrive in specific ecological niches.

15.5.2. Hybridization

The chosen donor species and the target crop plant are then hybridized using controlled hybridization techniques. These deliberate crossings are designed to produce hybrid offspring with genetic material from both parental species.

15.5.3. Backcrossing

The subsequent phase includes many rounds of backcrossing. During this step, the hybrid plants are mated with the target crop, systematically decreasing the genetic contribution of the donor species while conserving the genetic contribution of the target crop.

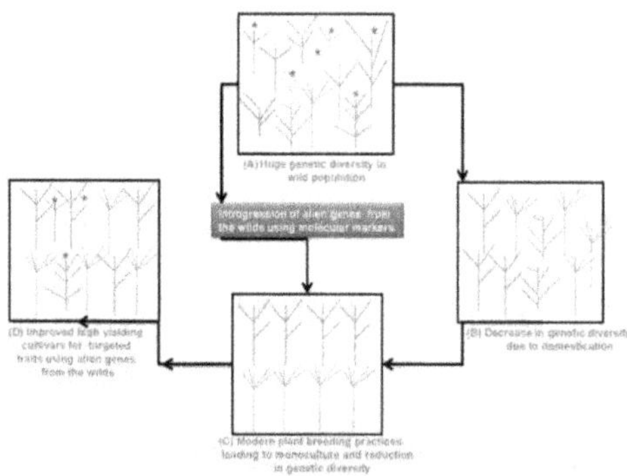

15.6. Cytogenetic Tools in Alien Gene Introgression

Identifying and tracing foreign genetic introgressions is critical in plant breeding, but this work can be difficult. To address these issues, genomic dot-blot hybridization has emerged as a beneficial approach, particularly in cases where standard molecular markers are difficult to utilize. This approach detects even the smallest genomic introgressions from species such as Hordeum chilense in the offspring of genetic crossings between wheat and H. chilense addition or substitution lines within the wheat genome.

In essence, genomic dot-blot hybridization is a critical tool for understanding and tracing the insertion of alien genes into the genetic makeup of a plant. Plant breeders can navigate the complexities of genetic introgressions with greater efficiency and precision by using cytogenetic approaches such as genomic dot-blot hybridization. This makes a significant contribution to the production of better and robust types of crops capable of meeting the changing needs of agriculture and food security.

15.7. Comparative Genomic Hybridization

Comparative Genomic Hybridization (CGH) is a molecular biology technique which enables the study for the comparison of DNA copy number variations between two different genomes. It is a valuable tool for detecting chromosomal abnormalities, gene amplifications, and deletions within the genomes under investigation.

15.7.1. Principle and Mechanism

CGH works on the premise of hybridizing DNA from two different organisms, typically a test sample and a reference sample, onto a microarray or other suitable platform. The test sample's DNA is tagged with one fluorescent dye (e.g., green), while the reference sample's DNA is labelled with another fluorescent dye (e.g., red). The microarray is then co-hybridized with the two labelled DNA samples.

15.7.2. The mechanism of CGH involves several key steps

- ❑ **DNA Labelling:** The test and reference samples' DNA is fragmented and tagged with fluorescent dyes.

- ❑ **Hybridization:** The labelled DNA from the test and reference samples are combined which are used in co-hybridizing onto a microarray with thousands of DNA probes. Each microarray probe corresponds to a certain genetic region.

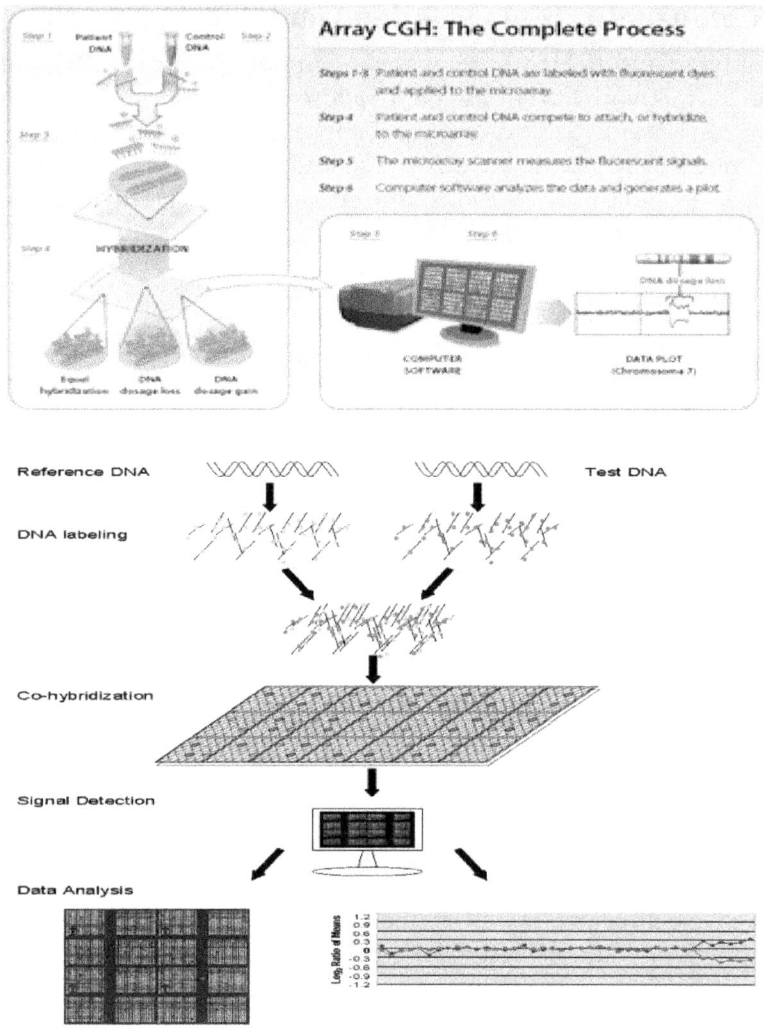

- [ ] **Fluorescence Detection:** The microarray is scanned for fluorescence signals after hybridization. The signal intensity represents the relative amount of DNA at each genomic site in the test and reference samples.

- [ ] **Data Analysis:** The fluorescence intensity data are examined to identify locations of the genome where the test and reference samples have different DNA copy numbers. When the test sample is compared to the reference sample, this analysis can identify chromosomal abnormalities such as deletions (loss of DNA) or amplifications (gain of DNA).

15.8. Applications in Crop Plants

- [ ] **Genomic Variation:** CGH technique was used to evaluate the genomic diversity among various crop varieties or accessions. Researchers can uncover genetic changes related with specific traits and tolerance to diverse environmental conditions by comparing the DNA copy number profiles of different kinds.

❏ **Stress Response:** CGH can help to unravel how crop plants respond to environmental stressors like drought or salinity. Researchers can find genes or genomic areas that have a role in stress tolerance by analysing copy number changes in stressed and non-stressed plants.

❏ **Disease Resistance:** CGH can be used to exploited the genetic basis of disease resistance in crop plants. Researchers can find genetic markers linked with resistance by comparing the genomes of resistant and susceptible plants, which will aid in marker-assisted breeding programs.

❏ **Chromosomal Aberrations Detection:** CGH is a useful method for determining chromosomal abnormalities such as translocations, duplications, or deletions that may affect crop production. This data was useful to drive breeding attempts to create crops with more stable and desirable genomes.

❏ **Breeding Programs:** In breeding programs, this tool was used to assess the genetic stability and integrity of crop types. It aids in the retention of desired features during the breeding process while minimizing unwanted genetic alterations.

In a comprehensive study utilizing Comparative Genomic Hybridization (CGH), researchers investigated the genomic structural variation between two maize inbred lines, B73 and Mo17, following thousands of years of maize domestication. The study revealed an astonishing level of structural diversity between these two genotypes, unmatched in higher eukaryotes. The analysis identified several hundred Copy Number Variations (CNVs) and several thousand Presence/Absence Variations (PAVs) within the genomes.

Specifically, CNVs are genomic regions where B73 and Mo17 have different number of DNA copies, reflecting substantial genetic diversity. Furthermore, PAVs, which are sequences present in one genome but entirely absent in the other, numbered in the thousands, with approximately 180 of them being single-copy genes. These gene-rich regions with variable presence or absence are likely crucial contributors to the extraordinary phenotypic diversity observed in maize. Overall, the study underscores the remarkable genomic diversity within maize, highlighting the crucial role in breeding program which is performed by structural changes in determining phenotypic differences between maize inbred lines and harnessing desirable traits in this agriculturally vital crop.

15.9. Somatic Hybridization

According to cytogenetics, somatic hybridization is defined as the process of developing hybrid organisms by fusing the protoplasts of two different plant species or varieties. It is a novel and efficient technique which involves the study of chromosomes, behaviour and genetic content, as leads to development of new crop varieties with improved traits. In this approach, protoplasts with two different genomes are necessary to fused before the desired somatic hybrid cells are determined and ultimately results in generation of hybrid plants (Evans DA and Bravo JE, 1988). The term protoplast refers to a cell that lacks cell wall either mechanically or enzymatically which constitutes a particular unique opportunity to generate cells with new genetic material and is also used in different species to overcome sexual compatibility.

15.9.1. Principle

Sexual hybridization has been employed to enhance the traits of cultivated plants in traditional method. The primary drawback of sexual hybridization is that it is only possible to produce offspring within a plant species or very closely related species, which also inhibits the advancements of new plant species. Somatic cell fusion that ultimately results in a viable hybrid may obtain over the species barriers for plant improvement that are present in sexual hybridization.

In terms of cytogenetics, somatic hybridization used in **karyotype analysis:**

Karyotype defined as the morphology of all the chromosomes in an organism (number) based on its their length, size, centromere position, secondary constriction, satellite, any differences between the sex chromosomes, and any other physical characteristics. The organization of chromosomes by size, shape, and banding patterns is called as Karyotyping. Somatic hybridization can produce hybrid cells with rearranged or mixed chromosome sets which can helpful in development of karyotypes for species. This information is valuable for taxonomic classification and studying chromosome evolution.

15.9.2. Steps in Karyotype Analysis

- ❏ **Culturing of cells:** Generally, the samples obtained from blood, tissue, cell culture, are stimulated to divide and typically involves culturing the cells in special culture medium under controlled conditions.

- ❏ **Mitotic Arrest:** This arrest is done at metaphase stage of mitosis, where actively dividing cells are treated with a chemical agent *viz.*, colchicine. As a result, the chromosomes are highly condensed and independent.

- ❏ **Cell Harvesting:** The chromosome containing nuclei are released after the arrested cells are then harvested and treated to disintegrate their cell membranes.

- ❏ **Chromosome Spreading:** The chromosomes are equally dispersed throughout the nuclei before they are placed on a glass slide. This is frequently accomplished by dropping a fixative onto the slide, which causes the chromosomes to disperse and the cells to rupture.

- ❏ **Staining:** The scattered chromosomes are stained with Giemsa stain or DAPI (4′,6-diamidino-2-phenylindole) to make the chromosome bands visible. These bands are unique patterns that aid in the identification of individual chromosomes and their structural abnormalities.

❏ **Microscopy and Analysis:** Under a microscope, the stained chromosomes are examined by the cytogeneticist or biologist to analyze the size, shape, and banding pattern of each chromosome to create the karyotype.

❏ **Karyotype Display: A** visual representation where the chromosomes are organized according to size and arranged in pairs based on homologous chromosomes in the form of image is known as karyotype display. For example, an average human karyotype has 23 pairs, with one chromosome from each pair derived from each parent.

The main purpose of karyotype analysis is to detect genetic disorders, cancer diagnosis and also useful in taxonomy and evolutionary studies of chromosomes.

15.10. Double Haploid

In cytogenetics, "double haploid" (DH) is described as an organism or cell that is homozygous at every genetic locus, which means that it contains two identical sets of each chromosome in its genome and it is determined by "doubling the haploid chromosome number". This technique mainly emphasis on chromosome study and also their organization within cells. This technique was very efficient and rapid tool to produce homozygous lines, cultivars, hybrids, genetic map construction and identification of molecular markers for selection of trait in crop improvement.

Usually, double haploids are developed when haploid cells, which have only one complete set of chromosomes, which is exactly half of the normal chromosome number (n) for the species.

15.10.1. Purpose of Double Haploids

❏ **Chromosome behaviour:** The chromosome behaviour is usually done at cell division *i.e.*, mitosis and meiosis are studied by using DHs. Their genetic homozygous makes it easier to analyse the chromosome segregation, recombination, and structural alterations in chromosomes.

❏ **Gene Mapping:** DHs are used in genetic mapping research to identify the positions of genes on chromosomes. In order to determine linkage relationships between genes and map their locations by using known genetic markers to generate DH populations.

❏ **Isolation of mutant:** DHs can be produced from individuals carrying mutations of interest. This enables the isolation of mutants with specific genetic changes, which is important for examining the function of gene and gene expression regulation.

❑ **Study of genetic stability:** DHs are useful in studying stability of genome and chromosome evolution because they have a stable genetic background over generations.

❑ **Genomics:** Both comparative and functional genomics uses DHs to understand the gene function and also used to compare the chromosomes of different organisms.

❑ **Genome Sequencing:** DHs are used to generate high-quality reference genomes for various organisms. These reference genomes serve as valuable resources for genome sequencing projects and the annotation of genes and regulatory elements.

❑ **Disease Studies:** DHs are used to study the genetic basis of diseases and disorders. By introducing disease-associated genetic variants into DHs, researchers can investigate the molecular mechanisms underlying these conditions.

❑ **Development of Marker:** Genetic markers like microsatellites or single nucleotide polymorphisms (SNPs) can be developed by using DH populations and are crucial for genetic studies and breeding programs.

15.11. Polyploid Manipulation

Polyploidy is a heritable condition of possessing more than two complete sets of chromosomes. According to the cytogenetics, polyploid manipulation is described as alteration of the ploidy level of an organism's cells or tissues. The polyploidy can be induced by chemical agents (colchicine) will interrupt the formation of spindle fibres during cell division which leads to doubling of chromosome; physical agents such as X-ray irradiation, heat shock etc., can damage DNA; hybridization (crossing different species with different chromosome numbers results in hybrid with different level of ploidy); invitro culture.

15.11.1. Significance of Polyploid Manipulation

❑ These polyploid crops often exhibit increased size, vigour, and disease resistance.

Examples: Include polyploid varieties of wheat, cotton, and strawberries.

❑ Polyploidy studies help in understanding genome evolution, genetic diversity, and the effects of chromosome doubling on traits.

❑ Polyploid cells can be used in regenerative medicine and tissue engineering to enhance cell and tissue growth.

❑ Inducing polyploidy in tumor cells can be used to study cancer progression and develop targeted therapies.

❑ In cases of endangered species with low genetic diversity, hybridization and polyploid manipulation can be used to generate individuals with higher genetic diversity, potentially improving their chances of survival.

❑ Polyploid cell lines can be engineered for the production of biopharmaceuticals.

❑ Polyploid microorganisms are explored for their potential in biofuel production due to their enhanced metabolic capabilities.

❑ In genomics, the polyploid manipulation is used to compare the gene expression and functional differences between different ploidy levels revealed the regulation of gene function in an organism.

Polyploidy manipulation in a population is analyzed by different techniques such as flow cytometry and karyotyping of chromosomes in a cell. It is powerful tool with a wide range of applications in agriculture, genetic research, medicine and biotechnology. It has practical potential for crop yield improvement, enhance the production of valuable products and also used in understanding of genetics and evolution of chromosomes in an organism.

15.12. Epigenetic Modification and Chromatin Remodelling

Chromatin is the complex of DNA and proteins that make up chromosomes, and it plays a crucial role in regulating gene expression. The structure of chromatin can be altered by a variety of mechanisms, including histone modifications and chromatin remodeling. So, as we discussed epigenetic modifications and chromatin remodelling are important processes that regulate gene expression in plants by modifying the chromatin status and recruiting transcription regulators. Histone modifications, such as acetylation, methylation, and phosphorylation, can alter the accessibility of DNA to transcription factors and RNA polymerase, thereby affecting gene expression. Chromatin remodelling, on the other hand, involves the repositioning or removal of nucleosomes, which can also affect gene expression. These complexes can be divided into several families, including the SWI/SNF, ISWI, and CHD families. Each family has a distinct structure and function, but they all share the ability to alter the structure of chromatin. The SWI/SNF family is the best characterized of the chromatin-remodeling complexes. It is composed of several subunits, including the ATPase subunit, which is responsible for the ATP-dependent remodeling of chromatin. The SWI/SNF complex can reposition nucleosomes, remove nucleosomes, or replace histones with variant histones. The ISWI family is another important chromatin-remodeling complex. It is involved in the repositioning of nucleosomes and the formation of nucleosome-free regions. The ISWI complex is also involved in the maintenance of chromatin structure during DNA replication and repair. The CHD family is involved in the regulation of gene expression by altering the structure of chromatin. It is involved in the formation of nucleosome-free regions and the repositioning of nucleosomes. These processes are crucial for plants to respond to salt stress and other environmental challenges. These processes are crucial for crop improvement through genome editing technologies. Chromatin remodelling involves the ATP-dependent repositioning of nucleosomes and alterations in the core histone composition of the nucleosome. Chromatin-remodelling complexes are responsible for these changes, which serve to regulate the accessibility of the genomic DNA, possibly in a locus-specific manner under specific conditions. Chromatin accessibility tends to decrease under salt stress in Arabidopsis thaliana, in contrast to the increase in chromatin accessibility under heat or cold treatment. Several chromatin-remodeling proteins in Arabidopsis thaliana have been implicated in salt stress response, including AtCHR12, AtCHR23, AtCHR2/BRM, and AtCHR3/SPLAYED (SYD). Knocking out AtCHR12, for example, abolished the inhibition on root growth by salt stress at low NaCl concentration.

15.13. Integrating Cytogenetics with Omics Technologies

Understanding the importance of integrating classic techniques of plant cytology with modern "omics" technologies is very crucial to advance our understanding of plant genomes and to develop new strategies for plant breeding and genetic improvement. From this integration first concept comes is the "chromosomics" which combines cytology with genomics (combination of chromosomes and cytology with gene content, structure and function for an entire organism), which was first proposed by Uwe Claussen. This understanding has led us to the development of two new concepts: plant cytogenomics and mutagenomics. The classical plant cytogenetics has now entered the functional genomic era, which has greatly facilitated the identification, localization, and mapping of chromosome-specific markers in plants at chromosome level. This is of high importance in plant breeding, molecular systematics, species identification, detection of hybrid nature, detection of alien chromosomes and chromosomal aberrations, and analysis of somaclonal variations and diversity. Mutation breeding is the technique where genetic information and tools are utilized in the designing of breeding strategies, screening, selection and verification/authentication of induced mutants, and desirable mutants are used for breeding process further. Some of the cytogenetics methods that have been integrated with omics technologies include DNA base-specific fluorescence banding, genomic in situ hybridization (GISH), fluorescence in situ hybridization (FISH), M-FISH and Pachytene-FISH, BAC-FISH, BAC landing-FISH, fiber-FISH, combined FISH, comparative FISH mapping, FISH on flow-sorted chromosomes, bar-code FISH, and primed in situ (PRINS) DNA labelling. The integration of cytogenetics methods and omics technologies has also led to the development of new techniques such as flow cytogenetics in chromosome functional biology, which is used to study the dynamics of chromatin architecture and cell cycle, representing chromosome functional biology. Overall, the integration of cytogenetics methods and omics technologies has greatly advanced our understanding of plant genomes and has led to the development of new strategies for plant breeding and genetic improvement. The sequencing projects oh humans were started initially by combining data of detailed physical maps, linkage analysis, somatic cell genetics and in situ hybridisation data. Even the data base for genome was also made with the help of chromosome-based genome assembly, dynamic chromatin states and several Some of the recent techniques and their uses are mentioned below

Cytogenetics has greater importance even in the genome projects where the basic requirements of genome sequencing can be fulfilled through several cytogenetic techniques as to know position of several orthologous sequences, reference individual/ population used in sequencing programme is free from chromosomal aberration and others. Basic karyotyping is also very much essential to understand the ploidy of individual before sequencing.

15.14. Digital Cytogenetics and Bioinformatics

Digital cytogenetics and bioinformatics play a crucial role in the field of genetics and genomics, revolutionizing how we study and understand the structure and function of genomes. These two are essential fields in modern biology and medicine, as they play a crucial role in understanding the genetic basis of diseases, evolution, and biodiversity. These disciplines combine advanced technologies and computational methods to analyze and interpret genetic information, offering numerous benefits. Digital cytogenetics and bioinformatics are increasingly important in the field of plant breeding, where they are used to enhance the efficiency and precision of breeding programs. These technologies help plant breeders select and develop improved plant varieties with desirable traits, such as increased yield, disease resistance, and adaptability to changing environmental conditions. Some major applications of these tools are described below:

Technique	Uses
Chromosome microdissection and sequencing	Sequencing individual chromosomes/ chromosome segment and haplotyping
Flow sorting of chromosomes and sequencing	Assigning sequences to chromosomes
Fibre-fluorescence in situ hybridisation (FISH)	Ordering of sequences large insert clones on a DNA fibre
Universal probe set and multiprobe slides	Rapid bacterial artificial chromosomes (BAC) mapping across multiple species
Homologue-specific oligopaints	Visually distinguish single copies regions of homologous chromosomes
Super-resolution microsocopy	Imaging of chromation and nuclear organisation
BioNano	Genomo mapping to improve assemblies and detect structural variants
Long-read sequencing (e.g., PacBio, Oxford Nanopore)	Improving genome assemblies, identifying structural variants
Linked-read sequencing (10X Chromium)	Phasing and improving scaffolding of genome assemblies
Hi-C sequencing and CHiA-PET	Chromatin interactions and improving genome assemblies (Hi-C)

Genome Analysis and Karyotyping: Digital cytogenetics and bioinformatics allow researchers to analyze entire genomes, identifying structural variations, gene mutations, and other genetic abnormalities. This is vital for understanding the genetic basis of diseases, such as cancer, and for developing targeted therapies. Digital cytogenetics is used to perform karyotyping digitally, identifying chromosomal abnormalities such as translocations, deletions, and duplications. This is vital for diagnosing genetic disorders like Down syndrome and leukaemia.

Genomic Selection and Marker-Assisted Breeding: Plant breeders use digital cytogenetics and bioinformatics to analyze the genetic makeup of plants. This information is crucial for identifying specific genes or genomic regions associated with desirable traits. Genomic selection allows breeders to predict the performance of plants based on their genetic profiles, accelerating the breeding process. Bioinformatics tools help design molecular markers (such as SSRs or SNPs) that are associated with important traits. These markers are then used to select plants with the desired traits more efficiently. These tools accelerate breeding Cycles by reducing the time required for breeding new plant varieties by enabling more precise selection of parental lines and by streamlining the evaluation of plant materials.

Disease Diagnosis and Treatment: Digital cytogenetics and bioinformatics are instrumental in identifying chromosomal abnormalities, gene mutations, and variations that underlie various genetic disorders and cancers. This knowledge enables early diagnosis and the development of targeted therapies. Digital cytogenetics and bioinformatics are instrumental in cancer research. They enable the identification of chromosomal abnormalities and driver mutations in cancer cells, helping researchers discover new therapeutic targets and biomarkers for early detection.

Evolutionary Studies: Bioinformatics and cytogenetics help researchers trace the evolutionary history of species and the genetic changes that have occurred over time. Comparative genomics reveals the shared genetic heritage among organisms and highlights adaptation to environmental challenges. Understanding the genetic diversity within and between species is vital for conservation efforts. DNA sequencing and bioinformatics enable the monitoring and preservation of endangered species and ecosystems which is very crucial for biodiversity conservation. These tools can also be used for several phylogenetic studies.

Genome Sequencing and Functional genomics: Bioinformatics is essential for processing and analysing NGS data, allowing researchers to study entire genomes. Digital cytogenetics and bioinformatics help identify structural variations like copy number variations (CNVs) and large-scale chromosomal rearrangements, which can be a possible cause for disease susceptibility. These tools are crucial for analysing gene expression data, understanding gene regulatory networks, and predicting gene functions as well as to study particular gene expression, regulation, and function. So, they have a major importance in the field of Functional genomics to study any specific trait.

Genetic Diversity and Germplasm Characterization: Digital cytogenetics and bioinformatics tools are used to assess genetic diversity within plant populations and germplasm collections. This information helps breeders identify valuable genetic resources for breeding programs.

Crop Improvement through Synthetic Biology: Bioinformatics plays a role in synthetic biology approaches for plant breeding. By designing and engineering genes or pathways, breeders can develop plants with novel traits or metabolic pathways.

Phenotyping and High-Throughput Data Analysis: Digital cytogenetics and bioinformatics are used to analyze phenotypic data, which includes various plant traits, at a large scale. Automated phenotyping platforms and data analysis tools assist in identifying promising plant varieties.

The use of digital image analysis and computer algorithms to automate the process of chromosome counting, which can accelerate the karyotyping process and make cytogenetic analysis quicker and less expensive which is a best example of digital cytogenetics. The automatic chromosome counting system uses a combination of Otsu's threshold selection method, hysteresis thresholding, median filtering, and thinning, along with newly developed methods for average width calculation, chromosome separation, and chromosome counting. The system comprises two main stages: pre-processing and counting. The pre-processing stage involves hysteresis thresholding, median filtering, thinning, separation, and cleaning, while the counting stage involves the use of a computer algorithm to automatically count the chromosomes in the pre-processed image. The system generally tested on a metaphase image database.

15.15. Genomic Sequence Data in Crop Cytogenetics

Genomic sequence data has emerged as a cornerstone in crop cytogenetics, providing unparalleled insights into crop plant genome structure, organization, and function. The combination of cytogenetics, the study of chromosomes, genomics, and also the organism's total genetic material, has transformed our ability to decode the genetic basis of crop attributes and use this information to improve crop yield.

15.15.1. Principle and Mechanism

Genomic sequencing is fundamentally concerned with finding the precise order of nucleotide bases (A, T, C, and G) within an organism's DNA. This method is based on the principles of DNA replication and nucleotide sequence detection. The technique differs depending on the sequencing technology employed, but the fundamental goal is to read and record the sequence of these bases.

- ❏ **Next-Generation Sequencing (NGS):** NGS technology, such as Illumina sequencing, dominate the field of genomic sequencing. The principle underlying NGS is parallel sequencing, in which millions of DNA fragments are sequenced at the same time. For example, in Illumina sequencing, DNA fragments are first sheared into shorter pieces. Each fragment is ligated with a unique adaptor before being amplified and placed onto a flow cell. The fragments are extended by inserting fluorescently labelled nucleotides, and the fluorescence released is measured. This technique is repeated iteratively, resulting in a large number of short sequences reads.

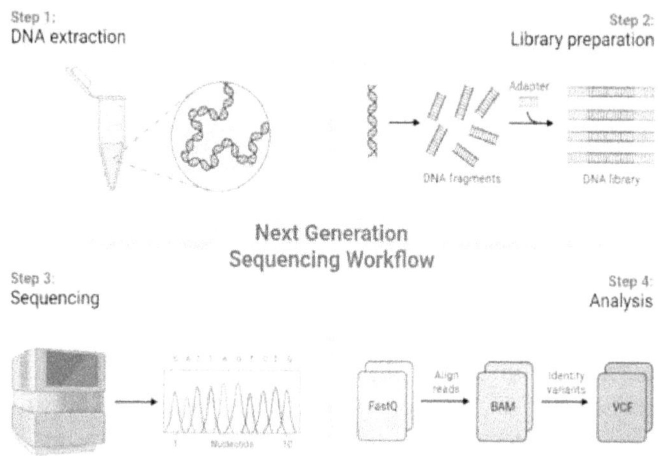

Fig 15.1. Next-Generation Sequencing

☐ **Third-Generation Sequencing (TGS):** TGS technology such as PacBio and Oxford Nanopore enable long-read sequencing. PacBio, for example, employs single-molecule, real-time (SMRT) sequencing. As DNA polymerase creates a complementary strand of DNA in real time, it produces light, which is captured as a kinetic trace. Nanopore sequencing involves passing DNA through nanopores and monitoring electrical current changes as nucleotides travel through. Both approaches generate long sequences, which are especially applicable for resolving complicated genomic regions.

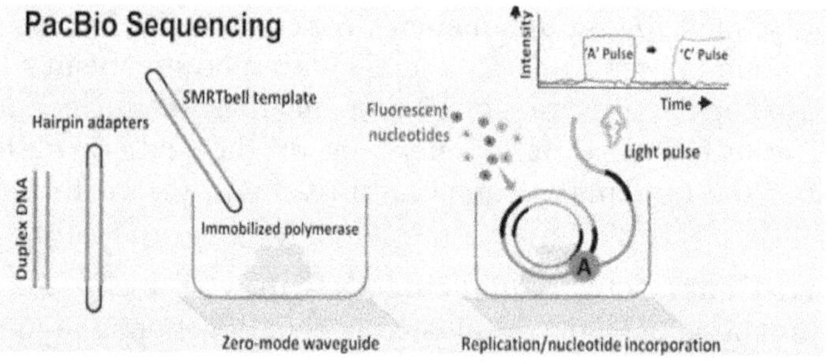

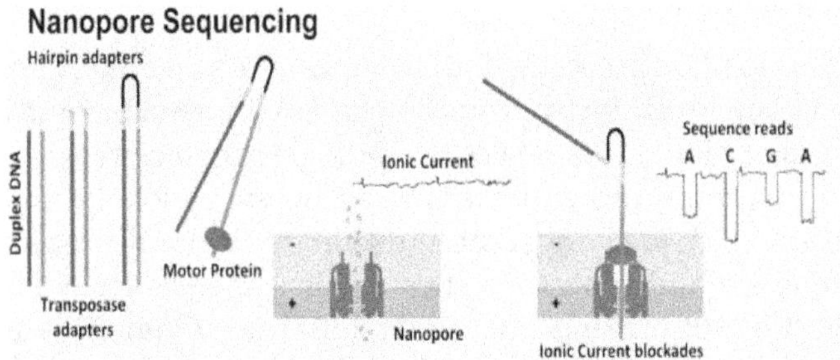

Fig 15.2. Third-Generation Sequencing

15.16. Applications in Crop Plants

☐ **Genome Assembly:** Genomic sequencing data is used to construct the whole genome of a crop plant. For instance, the genome of rice (*Oryza sativa*. L) was decoded using NGS methods. Understanding rice genetics and producing new varieties with improved traits have benefited greatly from the use of this reference genome.

☐ **Trait Mapping:** The genetic loci responsible for specific traits in crop plants are mapped using genomic data. Genomic sequence data was used to detect the quantitative trait loci (QTL) linked to traits like drought tolerance in maize (*Zea mays*). Breeders can create drought-resistant maize cultivars with the use of this data.

☐ **Precision Breeding:** The data of genomic sequence allows for precise and targeted breeding. In tomato (*Solanum lycopersicum*), researchers employed genetic markers identified through sequencing to create varieties with enhanced fruit flavour and shelf life.

☐ **Crop Improvement:** Genomic information aids in crop improvement by identifying genes associated with disease resistance. In wheat (*Triticum aestivum*), researchers observed that genes for resistance to wheat rust diseases, helping develop more robust and disease-resistant wheat varieties.

☐ **Genetic Diversity Preservation:** Genomic sequence data helps preserve genetic diversity within crop germplasm banks. The diversity of wild relatives of cultivated crops can be explored, as seen in wild soybean (*Glycine soja*), where genomic sequencing revealed valuable genetic resources for soybean improvement.

Crop cytogenetics is being transformed by genomic sequence data, which enables researchers and breeders to decipher crop genome mysteries, pinpoint desirable features, and quicken crop improvement efforts. By tackling the issues of food security and sustainable agricultural methods, this synergy between cytogenetics and genomics promises a better future for global agriculture.

15.17. Conclusion

With respect to plant breeding digital cytogenetics and bioinformatics have revolutionized plant breeding by providing tools to understand and manipulate the genetic basis of plant traits. These technologies enable breeders to develop improved crop varieties more efficiently, addressing global challenges such as food security, climate change, and sustainable agriculture. In summary, digital cytogenetics and bioinformatics are indispensable tools in biology and medicine, revolutionizing our understanding of genetics, disease, and evolution while paving the way for personalized medicine and conservation efforts. They continue to drive groundbreaking discoveries and innovations in the life sciences.

References

1. Schmidt C, Schindele P, Puchta H. From gene editing to genome engineering: restructuring plant chromosomes via CRISPR/Cas. *Abiotech.* 2020 Jan;1:21-31.

2. Schmidt C, Pacher M, Puchta H. Efficient induction of heritable inversions in plant genomes using the CRISPR/Cas system. *The Plant Journal.* 2019 May;98(4):577-89.

3. Park CY, Sung JJ, Kim DW. Genome editing of structural variations: modeling and gene correction. *Trends in Biotechnology.* 2016 Jul 1;34(7):548-61.

4. Ara A, Sofi PA, Rather MA, Dar ZA, Bhat MA, Hamid A. Alien Gene Introgression in Crop Improvement–New Insights. *Int. J. Pure App. Biosci.* 2018;6(6):890-904.

5. Rey MD, Prieto P. Detection of alien genetic introgressions in bread wheat using dot-blot genomic hybridisation. *Molecular Breeding.* 2017 Mar;37(3):32.

6. Mu H, Wang B, Yuan F. Bioinformatics in Plant Breeding and Research on Disease Resistance. *Plants.* 2022 Nov 15;11(22):3118.

7.Springer NM, Ying K, Fu Y, Ji T, Yeh CT, Jia Y, Wu W, Richmond T, Kitzman J, Rosenbaum H, Iniguez AL. Maize inbreds exhibit high levels of copy number variation (CNV) and presence/absence variation (PAV) in genome content. *PLoS genetics.* 2009 Nov 20;5(11):e1000734.

8. O'Connor, C. (2008) Fluorescence in situ hybridization (FISH). Nature Education 1(1):171 9.https://www.ncbi.nlm.nih.gov/pmc/articles/PMC7122835/#:~:text=Fluorescence%20in%20situ%20hybridization%20(FISH)%20is%20the%20most%20convincing%20technique,to%20various%20types%20of%20cancers

10. Touchell, Darren H., Irene E. Palmer, and Thomas G. Ranney. "In vitro ploidy manipulation for crop improvement." *Frontiers in Plant Science* 11 (2020): 722.

11.Waara, Sylvia, and Kristina Glimelius. "The potential of somatic hybridization in crop breeding." *Euphytica* 85 (1995): 217-233.

12. Shuro, Aliyi Robsa. "Review paper on the role of somatic hybridization in crop improvement." *International Journal of Research* 4, no. 9 (2018): 1-8.

13. Deakin, J. E., Potter, S. and O'Neill et al. (2019). Chromosomics: Bridging the gap between genomes and chromosomes. *Genes, 10*(8): 627.

14. Silva, G. S., and Souza, M. M. (2013). Genomic in situ hybridization in plants. *Genet Mol Res, 12*(3): 2953-2965.

15. Yung, W. S., Li, M. W., Sze, C. C., Wang, Q., and Lam, H. M. (2021). Histone modifications and chromatin remodelling in plants in response to salt stress. *Physiologia Plantarum, 173*(4): 1495-1513.

16. Talukdar, D., and Sinjushin, A. (2015). Cytogenomics and mutagenomics in plant functional biology and breeding. *PlantOmics: The omics of plant science*, 113-156.

17. Gajendran, V., and Rodríguez, J. J. (2004). Chromosome counting via digital image analysis. In *2004 International Conference on Image Processing, 2004. ICIP'04.* (Vol. 5, pp. 2929-2932). IEEE.

18. Bharti, R. and Grimm, D.G., 2021. Current challenges and best-practice protocols for microbiome analysis. *Briefings in bioinformatics, 22*(1), pp.178-193.

19. Brozynska M, Furtado A, Henry RJ. Genomics of crop wild relatives: expanding the gene pool for crop improvement. Plant biotechnology journal. 2016 Apr;14(4):1070-85.

20. Edger PP, Poorten TJ, VanBuren R, Hardigan MA, Colle M, McKain MR, Smith RD, Teresi SJ, Nelson AD, Wai CM, Alger EI. Origin and evolution of the octoploid strawberry genome. Nature genetics. 2019 Mar;51(3):541-7.

21. Huang X, Yang S, Gong J, Zhao Y, Feng Q, Gong H, Li W, Zhan Q, Cheng B, Xia J, Chen N. Genomic analysis of hybrid rice varieties reveals numerous superior alleles that contribute to heterosis. Nature communications. 2015 Feb 5;6(1):6258.

22. Jiao Y, Peluso P, Shi J, Liang T, Stitzer MC, Wang B, Campbell MS, Stein JC, Wei X, Chin CS, Guill K. Improved maize reference genome with single-molecule technologies. Nature. 2017 Jun 22;546(7659):524-7.

23. Project IR. The map-based sequence of the rice genome. Nature. 2005 Aug 11;436(7052):793-800.

24. Tello-Ruiz MK, Naithani S, Stein JC, Gupta P, Campbell M, Olson A, Wei S, Preece J, Geniza MJ, Jiao Y, Lee YK. Gramene 2018: unifying comparative genomics and pathway resources for plant research. Nucleic acids research. 2018 Jan 4;46(D1):D1181-9.

25. Tomato Genome Consortium X. The tomato genome sequence provides insights into fleshy fruit evolution. Nature. 2012 May 30;485(7400):635.

Chapter – 16

Seed Systems for Rapid Adaptation of Improved Crop Varieties

Bangi Sharatchandra, Rashmi Jha, and Anu

Ph.D. Scholars, Department of Seed Science and Technology, TNAU

16.1. Introduction

Nearly 70% of India's population depends on agriculture and related industries for their livelihood and survival. It is an agrarian nation with a variety of climate zones. The percentage of India's GDP that comes from agriculture is between 15 and 17 percent. According to the "India Rural Development Report 2012-13" compiled by the IDFC Rural Development Network, India is the largest and fastest-growing country, but a significant portion of its arable land is under the control of marginal and small farmers. At the moment, land holdings are too small to generate adequate household income, but these small farms are more productive, especially when growing crops that require intensive labor and livestock management. The government promotes the adoption of cutting-edge technological techniques including multiple cropping, intercropping, and integrated farming systems, among others, to increase crop output. In order to improve the situation of Small and Marginal Farmers by doubling their income by 2022, the government is reorienting its interventions from a production-centric strategy to farmer's revenue-centric initiatives, with an emphasis on better and new technology advancements

16.2. History Of Quality Seed Production

The idea of generating high-quality seeds gained traction with the establishment of the Swedish Seed Association in 1886, which marked the beginning of organized seed production and distribution. The primary goals of this association were the production and distribution of forage crop seeds. Canadian scientist Dr. JW Robertson proposed the production of foundation seed in 1917, and the International Crop Improvement Association (ICIA) was founded in 1919 to oversee the development of guidelines and standards for the production and certification of high-quality seeds. The association was renamed the Association of Seed Certification Agencies (AOSCA) in 1969. In 1946, ICIA created four seed classifications specifically for fodder crops. In 1968, these classifications were extended to additional grain crops.

The Uttar Pradesh Department of Agriculture in India began producing and disseminating 150 tons of wheat seed in 1900, but there were only a few basic seed testing facilities in Kanpur. In 1920, the Uttar Pradesh government began making efforts to construct seed storage facilities in tehsils and subdivisions and emphasized the importance of producing and disseminating high-quality seeds. According to Sahu et al. (2018), the Royal Commission on Agriculture looked into the production and distribution of seeds in India in 1925.

Private seed companies made their debut around 1945 with examples like Sutton's for vegetables grown in temperate climates. Independent seed growers established the All India Seed Producer's Association (AISPA) in 1946. As stated by the Grow-More Food Program Committee in 1952 and the Famine Enquiry Commission in 1945, high-quality seeds of improved varieties must be multiplied and distributed. A formal seed system was therefore developed for India as part of the Second Five Year Plan (1956–1961) in order to produce nucleus and breeder seeds, which were used to multiply other classes of seeds (Sahu, et al., 2018).

Programs to produce wheat and maize seeds as well as rice seeds were introduced in the years 1966–1967 and 1967–1968, respectively. Following a thorough evaluation and recommendations, the Indian Seed Act became operative on October 2, 1969. The Indian Seed Act of 1966 aims to regulate both the quality of particular seeds sold for sale and related issues. This act's key characteristics are as follows:

- ❑ To advise the federal and state governments on the Act, the Indian government established the Central Seed Committee.
- ❑ The formation of the Central Seed Laboratory.
- ❑ Establishing state-level seed labs to assess the quality of seeds.
- ❑ Notification of variations by the Indian government.
- ❑ Minimum standards have been set for seed germination and quality, and labeling must be done.

- ❑ Developing seed standards.
- ❑ Assure that the seeds may be recognized as identity of the proclaimed variety.
- ❑ The requirement of a minimal standard of purity and germination.
- ❑ Demanding the correct labels on seed containers.
- ❑ The setting up of agencies for seed certification.
- ❑ The formation of the Central Seed Certification Board.

This explains the development of India's institutional seed system over time, as well as the country's regulatory structure and ongoing commitment to the production and distribution of high-quality seeds.

16.3. The Role of Quality Seed Production in India

Seed, a crucial transporter, carries the genetic potential of many crop types. Farmers' ability to reap excellent yields and financial rewards depends on timely availability to top-notch seeds. Effective certification procedures, which have evolved significantly in India's seed sector, are essential for producing high-quality seeds. The public sector historically controlled the Indian seed industry until the 1980s, but it has gradually changed into a diverse sector with about 500 seed enterprises and firms. This transformation has been fueled by two factors: an increased focus on R&D initiatives and the liberalization of governmental policies (Chauhan *et al.*, 2016).

At all pertinent levels—local, regional, and international—seed is managed by bilateral and multilateral agreements as a traded good. These agreements place an emphasis on seed quality rather than quantity. The Indian Minimum Seed Certification Standards (IMSCS), which are administered by the Seeds Act of 1966, stipulate that high-quality seeds in India must adhere to these standards.The International Seed Testing Association (ISTA) in Switzerland, the Organization for Economic Co-operation and Development (OECD) Seed Schemes in France, and the Association of Official Seed Analysts (AOSA) in the USA must coordinate testing and certification procedures between domestic and international organizations. This harmonization is necessary for the development of standards and rules that enable the Indian seed industry to successfully compete in the seed trade (Prasad, *et al.*).

Notably, ISTA and OECD guidelines and the present seed quality standards and regulatory frameworks in India differ from each other. Both Tonapi et al. (2014) and Kamble et al. (2015) note that there are variations in the types and specifications of certificates, the range of crops, the minimum lot sizes, standard procedures for physical purity analysis, viability testing, seed health testing, testing seed with weighed replicates, calculating seed quantities by number, and seed vigour testing, as well as methods and standards for seed crops. Since joining the OECD, India has made great strides toward standardizing seed quality (Trivedi, 2012). It also offers suggestions for ways to restructure the seed production chain in order to successfully address the ongoing and emerging problems with the production of high-quality field crop seeds.

Pure seeds are an essential component for wholesome crops and robust production. The seeds must be free of impurities and meet the minimum recommended quality criteria. To safeguard the quality of seeds for Indian farmers, India's seed production system recognizes a number of seed classes, including Nucleus, Breeder, Foundation, and Certified seeds. Each of these seed classes has its own set of quality requirements. The timely availability of high-quality seeds created to meet specific agricultural requirements is a vital element in increasing grain production. Small and marginal farmers in India have significant challenges in their farming because of a lack of resources, particularly access to high-quality seeds. A robust and active system of seed production and distribution is therefore essential to the country's food security and agricultural development. Seeds are of utmost importance in agriculture since they serve as the cornerstone of agricultural sustainability. In an effort to ensure seed quality, accessibility, and availability during the past seven decades, the seed system has been upgraded.

16.4. Regulation of Seed System In India

The two main categories of the seed system in India are formal and informal. These systems are necessary to guarantee that farmers across the country have access to high-quality seeds. These initiatives seek to grow the unorganized seed market, close the seed quality gap, and speed farmers' adoption of superior crop types.

16.4.1. Formal Seed System

The formal seed system is characterized by a structured chain of activities leading to clearly defined seed products. It is primarily composed of public sector research institutions, public and private sector organizations involved in seed production and marketing, and organizations in charge of seed certification and quality control. It is a deliberate, well-thought-out procedure that often begins with the development of various varieties and hybrids. One of the core principles of the Formal Seed System (FSS) is to maintain varietal integrity and uniqueness while guaranteeing the production of seeds with the highest level of physical, physiological, and sanitary quality (Reddy et al., 2007). The official seed system is managed by government entities (such as government institutions, state-owned farms, university farmland, and Krishi Vigyan Kendras), registered seed producers, including non-governmental organizations (NGOs), and private companies (Sahu *et al.*, 2018; Prasad *et al.*, 2017).

Formal Seed System: Started in India in 1964, the formal seed system is governed by a variety of committees and regulations. It entails a sequence of steps that lead to the introduction of crop varieties that have received official approval.

- ❏ The Central Variety Release Committee (CVRC), a federal organization, is in charge of controlling the release of crop varieties.

- ❏ The State Variety Release Committee (SVRC), a state-level organization, is in responsibility of regulating variety releases.

- ❏ The Seeds Act of 1966: The Indian government enacted this law to regulate seed quality and certification.

- ❏ The Central Seed Committee (CSC), which replaced the CVRC, provides guidance to the federal and state governments on matters relating to seeds

The formal system gives top priority to maintaining genetic integrity, cultivating seeds with the highest level of hygienic, physiological, and physical quality, and producing seeds with a distinctive varietal identity. Variety certification under the official seed production program is only possible after they have received formal notification.

Breeder seed production: It is essential for establishing the seed production process and raising awareness of recently introduced varieties. Different groups and companies can order breeder seed based on their requirements, and it will then be produced under control. Even though breeder seed does not go through seed certification, it is nevertheless under the supervision of a diversified staff.

Foundation and Certified Seed Production: Foundation seed is mostly produced at the farms of organizations that produce seeds, and it goes through a number of phases of multiplication. On the other hand, certified seed is grown in farmers' fields under the constant observation of seed-producing organizations.

16.4.2. Informal Seed System

However, the informal seed system (ISS), often known as the village, farmer, or local seed system, operates in a different manner. Under this system, farmers have direct access to, and a say in, the production, distribution, and storage of seeds from their own crops. Additionally, they would exchange seeds with friends, neighbors, relatives, or vendors at local grain markets. The variations passed down through the informal system typically contain landraces or mixed races, creating heterogeneous combinations of various sorts. Both the formal and informal seed systems have challenges that must be addressed creatively.

Informal Seed Systems: The informal seed system, also known as the farmer, village, or local seed system, is one in which farmers directly create, share, and get seeds. A heterogeneous mixture of various types results from this system's reliance on commerce between farmers, neighborhood markets, and occasionally landrace or mixed races.

Developing the Informal Seed Industry: In India, there is a significant gap between the supply of certified seed available through the legal seed system and the demand for it. A number of strategies have been proposed to address this problem and encourage the spread of improved cultivars.

- ❑ **Seed Village Program:** This initiative provides foundation seed and technical support to interested farmers so they can begin village-level seed production. Farmers are permitted to freely trade their harvested seed with other farmers within or outside of the hamlet.

- ❑ **"Beej Swavlamban Yojna" (A Scheme for Building Self Reliance in Seed):** As part of this initiative, farmers are provided with a small amount of foundation seed that is carefully inspected by agricultural inspectors. The remaining seed is used for commercial crops, and some of it is recycled to **keep the seed production process going.**

- ❑ **Contract Seed development:** Local seed vendors may promote the development of high-quality seeds by hiring specific farmers to cultivate improved varieties. Technical knowledge is crucial for both farmers and dealers.

Contrarily, the informal seed system (ISS) is primarily managed by farmers themselves, with sporadic support from private seed growers. The ISS lacks the formal quality control framework and procedures present in the FSS. One of the challenges with the ISS is the potential for varietal deterioration when the same variety is repeatedly replicated year after year. This deterioration can lead to genetic drift, random mutations, unintentional pollination or outcrossing, and seed mixing. These factors combined may have an impact on crop performance and varietal genetic purity. The Nucleus, Breeder, Foundation, and Certified seed categories, each of which has certain quality requirements, were developed by the FSS in contrast to address these difficulties. However, these safeguards are frequently absent from the ISS.

In order to solve this issue, farmers and seed growers must be made aware of the importance of producing high-quality seeds in their fields for personal use. More than 85% of the seeds used in India today are produced by farmers, yet only 12% of these seeds meet the criteria for high-quality seed (Reddy et al., 2007). A 10–20% decline in crop productivity is attributable to the variation in seed quality.

As a result, both the formal and informal parts of India's seed system have unique management structures and quality control processes. The informal system, which is dominated by farmer-saved seeds, frequently lacks these safeguards, which leads to varietal degradation and decreased crop yields. In contrast, the formal system uses specific seed categories to assure seed quality. Farmers and seed growers must be made more aware of their responsibilities in order to improve seed quality in the unregulated market (Sahu *et al.*2018).

16.5 Establishment of Central Varietal Release Committee (CVRC); State Varietal Release Committee (SVRC) and Central Seed Committee (CSC)

One cannot overstate the value of seed in agriculture. The main difficulty in the Indian subcontinent up until the 1960s was the creation of high-quality seeds.

Prior to 1960, there was no high-quality seed production in India, and as farmers were unaware of the proper postharvest practices, they opted to use farm-stored seeds instead. After the initial phase of the green revolution initiative in India began, the value of high-quality, high-yielding hybrid and varietal seeds became apparent. The Essential Commodities Act of 1955 designated seed as an essential commodity in recognition of the need of high yielding, high-quality seeds. for ensuring that a bigger portion of the farming community can access them. The Government of India (GOI) established specific protocols for the official release of varieties through the Central and State Varietal Release Committee (CVRC) and State Varietal Release Committee (SVRC) systems in order to make them accessible to a larger portion of the farmer community. This was done back in October 1964. Later, the Indian government created The Indian Seeds Act, 1966, which allowed for the establishment of the Central Seed Committee (CSC) in 1969 and kept it under the supervision of the Ministry of Agriculture, Cooperation, and Farmers Welfare to carry out the duties of the CVRC for ensuring the sale and notifying of the quality of seeds of the kinds/varieties. The State Seed Sub-Committee (SSSC) was established to carry out similar duties at the state level, and the CSC established a Central Sub Committee on Crop Standards, Notification & Release of Varieties for Agricultural Crops and Horticultural Crops to perform the function of cultivar release/notification, provisional notification, and de-notification.

16.6. Development of a Variety and Performance Assessment Trials Followed

Plant breeders and organizations create entries (pure lines, open pollinated varieties, and hybrids) for enhancing humans (food and nutritional security) by scientific methodologies and extensive breeding operations. Many organizations (ICAR or non-ICAR national institutes, SAUs, private national or multinational companies, etc.) employ both traditional (introduction, selection, hybridization followed by selection, etc.) and cutting-edge (tissue culture-based techniques such as anther culture, somaclonal variation, and pollen culture; marker assisted breeding, mutation, etc.) breeding methods to produce elite material with high yield potential. Prior to incorporating superior cultivars into their breeding program, the concerned plant breeders repeat their three- to four-year stability tests on the created elite materials at their research station. Plant breeders and organizations use scientific techniques and extensive breeding programs to develop entries (pure lines/open pollinated varieties) for the benefit of humanity in the areas of food security and nutrition. Many institutions (ICAR or non-ICAR national institutes, SAUs, private national or multinational companies, etc.) use both conventional (introduction, selection, hybridization followed by selection, etc.) and cutting-edge (tissue culture-based methodologies such as anther culture, somaclonal variation, and pollen culture; transgenic or genome editing techniques, marker assisted breeding, mutation, etc.) breeding approaches to produce elite material with high yield potential. AICCIP trials enable additional testing of improved cultivars in numerous settings around the country. The elite materials developed by the concerned plant breeder(s) are examined for stability over a period of three to four years in replications at their research station. The following prerequisites must be met for newly submitted material or admission into the three-tier system:

- ❐ **Station trial or preliminary yield trial:** The responsible plant breeder must carry out a station or regional trial, and the proposed entry must have successfully passed a stringent process of evaluation or screening (insect pests and diseases). Tolerance to abiotic stressors and crop-based quality standards must both be determined to be necessary. The PC/PD (project coordinator / project director) need access to pre-coordinated trial data on yield, trait stability, and other relevant agronomic traits to validate the entry's applicability.

- ❐ The submission must meet the fundamental standards for seed certification with regard to physical purity, germination rate, phenotypic uniformity, and genotypic stability. The entry must possess just a few unique qualities to set it distinct from the other types that are still in use. These differentiating traits aid in the identification of variety during legal infringement (DUS testing).

- ❐ Any information on the parentage or pedigree of the entry, such as that provided by the involved plant breeder or agency, should be available to the PC/PD. The variety produced by using a diversity of parents in the breeding program will be favoured if entries perform equally well in the coordinated trials.

- ❐ Private businesses can submit their materials to the coordinated trial system in the same manner as other organizations, but they must comply with Government of India regulations and pay the set entry cost. Every material used, including synthetics, composites, hybrids, and products of selection, hybridization followed by selection, etc., must go through the same system.

Next to this a three-tiered system of varietal trial comes i.e., initial varietal trial (IVT), advanced varietal trial I and II (AVT I & II). These are conducted in AICCIP centers for a variety of crops, which are found at ICAR institutes, State Agricultural Universities (SAUs), or other volunteer centers chosen by the organization. Superior entries from IVT will be allowed for the trial in the case of AVT-I based on a 5–10% improvement over the top performing check. Due to the smaller plot size in AVT-I than in IVT, it will have fewer entries that have been tested than IVT, and the data on yield and other auxiliary features will be more accurate, trustworthy, and error-free. In compared to IVT, more testing locations should be used to provide more accurate data on yield and other economically significant features, varietal adaptation, biotic and abiotic tolerance, quality metrics, and other factors. National, zonal, and local checks (as used in IVT) will be used in addition to the entries for critical inspection.

Based on their performance over the top check-in performers in each zone, the most promising entries would be admitted to the AVT-II.

Similar to AVT-I, all prerequisites for Advanced Varietal Trial II (AVT-II) must be met. However, at the AVT-II stage, additional data will be generated, such as responses of entries to different sowing dates, seed rates, plant and row spacing (population density), behavior in various levels of fertilizer and irrigation, responses of diseases and pests by sponsored plant pathologists, and crop quality parameters by biochemists. The seed technology institute will develop descriptors to aid with seed certification. From cooperating and volunteer locations/centers, the program coordinator must get all processed and evaluated data on yield and other pertinent features. Annual reports for each crop are produced based on these numbers.

At a predetermined institute/university's annual crop workshop or national group gathering, the top test entries will be picked based on their performance over the previous three years. At the national group meeting, the Principal Investigators and Zonal Coordinators present more details about the varieties. With the agreement of the Indian Council of Agricultural Research (ICAR) Deputy Director General (Crop Science), a "Varietal Identification committee (VIC)" was established prior to an annual workshop or national group meeting.

The VIC supplies the Central Subcommittee on Crop Standards, Notification, and Release with the relevant information regarding authorized entries. This committee has the exclusive authority to release and notify the top entries at national or zonal levels based on the recommendations of the VIC. In accordance with Section 3 of the Seed Act of 1966, the Central Seed Committee established a Central Sub-Committee on Crop Standards, Notification, and Varieties Release in 1994. There is one chairman and seventeen members on the committee. While the State Seed Sub-Committee publishes varieties for particular states, the Central Sub-Committee releases varieties for the benefit of stakeholders and to address regional, zonal, or national needs. Notification regularly permits certified seed production by domestic public or private seed multiplication organizations. The varieties/hybrids will be issued for the appropriate agro-climatic zones, which may encompass one or more states or the entire country, when the Central Sub-Committee approves the application.

16.7. Release of Varieties or Cultivars

The intent of cultivar release is to publicize newly developed cultivars to a larger audience for widespread cultivation in climate-appropriate areas. It aids farmers in choosing cultivars for a certain geographic region. In other words, the introduction of a cultivar is a suggestion to farmers to use it. Variety release is thus linked to variety notice even if the release procedure does not have any statutory protections. It is necessary to bring the seed of a certain crop variety under the notification system since only notified varieties will be subject to Seed Law Enforcement. Only varieties that have been formally notified may be used as samples by a seed inspector for testing and seed quality assurance. Before the Gazette of India issues a notification, a released variety cannot enter the seed chain. These concerns will make the notification necessary for other things to be done.

On the recommendation of the Central Seed Committee, the Central Government issues the notification. In order to ensure that farmers receive high-quality seeds, notification is necessary for the production of certified seed. Variety becomes an asset for the Indian government after notification. When a specific variety has received formal notification and only those varieties enter the seed supply chain, breeder seed can be produced. In addition to regulating any infringement involved in the later stages of varietal promotion, notification also contributes to the emergence of original varieties depending on their genealogy.

The following are the key differences between varieties that have been released and those that have been notified: (Source from Chand *et al.*, 2020)

Released variety	Notified variety
Under the Seed Act of 1966, it is not an official duty.	Section 5 of the Seed Act of 1966 requires the registration of statutory function and variety.
It is ineffective for use in seed certification.	Seed certification will only apply to varieties that have been informed.
Farmers have no assurances on seed quality.	Quality seeds that farmers can rely on.
Samples of seeds cannot be drawn and tested by seed enforcement organizations (such as seed inspectors).	They are allowed to collect and examine seed samples.
These are not Indian government property.	The Government of India owns the Notified varieties.
Its primary goal is to make public information about the cultivar and its adoption area available.	Regulation of seed quality is the primary goal.
It is challenging to identify the origin	It will be easier to determine its origin with notification of the varieties of crops.

16.8. Notification of a Variety

Breeder, foundation, and certified classes of seeds, as well as the certified seeds themselves, cannot be produced without variety notice. Variety notification, which is carried out by the certification procedure, is an essential part of the seed supply chain. Such a variety must be informed before the seeds from that variety can be certified.

The Seeds Act mandates that only varieties that have been notified may be used to produce certified seeds, placing notification ahead of certification (as opposed to the appropriately labelled seed class). There must be notification for the creation of certified seeds. The variety must be told before the seed may be accepted. A seed law enforcement agency is permitted to gather and examine samples of seeds from species that have been notified under the Seeds Act. As a result, notification is necessary for seed quality regulation. Farmers benefit from variety notification because they can get guaranteed-quality seeds of the notified varieties. The Seeds Act states that seeds of notified kinds may be sold after being appropriately packaged and labeled to state the minimum requirements. A variety usually belongs to the Indian government when it is reported. The Central Seed Committee keeps morphological records of the registered cultivars in order to combat biopiracy. The variety's disclosure will help identify their source. Subsidies are being assessed based on the notification status. Program activities and seed planning are being carried out using the notification statistics data.

16.9. Variety Denotification

If a species has been notified but has not performed well in the chosen area, has been grown for more than 15 years, or is not in high demand, it may be denotified. The Indian government has the right to denotify in accordance with the recommendations of the central seed committee. A vital component of agricultural management and plant breeding, release, notification, and denotification of novel plant varieties ensures that farmers have access to the most cutting-edge, productive plant varieties to boost crop yields, quality, and resilience. To protect the integrity of the plant breeding industry, the agricultural authorities frequently regulate and watch on these activities.

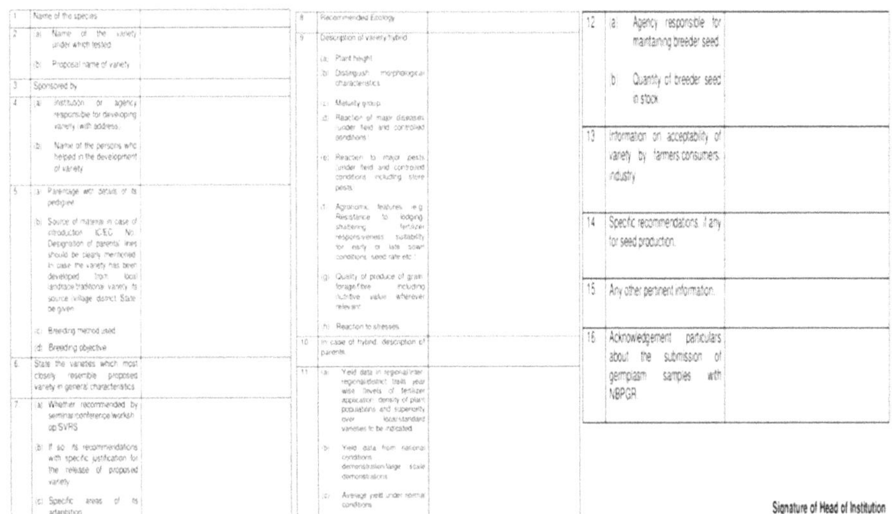

1		Name of the species	
2	(a)	Name of the variety under which tested	
	(b)	Proposal name of variety	
3		Sponsored by	
4	(a)	Institution or agency responsible for developing variety (with address)	
	(b)	Name of the persons who helped in the development of variety	
5	(a)	Parentage and details of its pedigree	
	(b)	Source of material in case of introduction IC/EC No. Designation of parental lines should be clearly mentioned in case the variety has been developed from local landrace/traditional variety its source (village, district, State) be given	
	(c)	Breeding method used	
	(d)	Breeding objective	
6		State the varieties which most closely resemble proposed variety in general characteristics	
7	(a)	Whether recommended by seminar/conference/workshop/SVRS	
	(b)	If so its recommendations with specific justification for the release of proposed variety	
	(c)	Specific areas of its adaptation	
8		Recommended Ecology	
9		Description of variety/hybrid	
	(a)	Plant height	
	(b)	Distinguish morphological characteristics	
	(c)	Maturity group	
	(d)	Reaction of major diseases under field and controlled conditions	
	(e)	Reaction to major pests under field and controlled conditions including store pests	
	(f)	Agronomic features e.g. Resistance to lodging, shattering, fertilizer responsiveness, suitability for early or late sown conditions, seed rate etc.	
	(g)	Quality of produce of grain/forage/fibre including nutritive value wherever relevant	
	(h)	Reaction to stresses	
10		In case of hybrid, description of parents	
11	(a)	Yield data in regional/inter-regional/district trials year wise (levels of fertilizer application, density of plant populations and superiority over local/standard varieties to be indicated)	
	(b)	Yield data from national conditions/demonstration/large scale demonstrations	
	(c)	Average yield under normal conditions	
12	(a)	Agency responsible for maintaining breeder seed	
	(b)	Quantity of breeder seed in stock	
13		Information on acceptability of variety by farmers/consumers/industry	
14		Specific recommendations if any for seed production.	
15		Any other pertinent information.	
16		Acknowledgement particulars about the submission of germplasm samples with NBPGR	

Signature of Head of Institution

Fig.1: Proforma for Submission of Proposal of Release of Crop Verify to Central Sub-Committee on Crop Standards, Notification and Release of Varieties (Central Varieties)

Fig.2: Proforma for Submission of the Proposal for Notification of Crop Varieties of Vegetable Crops Under Section 5 of the Seeds Act 1966 (State Varieties)

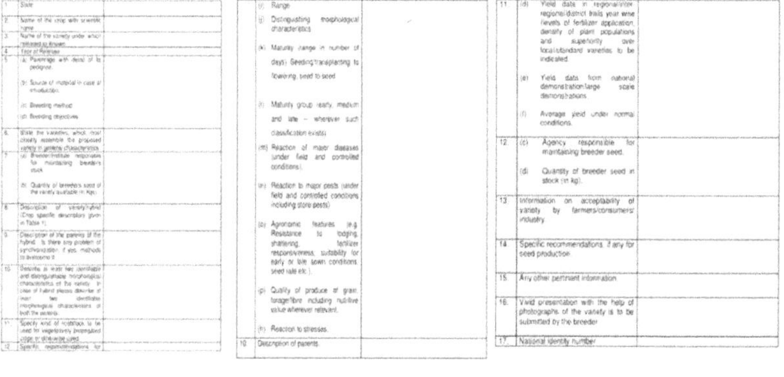

Fig.3: Proforma for Submission of Proposal of Release of Crop Variety to State Seed Sub Committee

PROFORMA FOR DENOTIFICATION OF CROP/VARIETIES UNDER SECTION 5 OF THE SEEDS ACT, 1966.

1. State

2. Crop

3. Variety

4. Year of Notification and S.O No.

5. Name of the Organising Breeder/Institute of the variety

6. Reason for denotification

7. Name of the variety which has replaced/will replace the denotified variety.

8. Recommendation of the All India Workshop about the variety

9. Acknowledgement particulars about the germplasm samples with NBPGR.

Signature of the Head of the Institute/
Co-convenor of the State Seed Sub Committee.

Fig.4: Proforma for Denotification of Crop/Varieties Under Section 5 of the Seeds Act, 1966.

The survival and reproduction of plant species from one generation to the next depend on seeds, which are exceptional genetic storage units. They can grow into identical plants and are the fundamental building blocks of reproduction in flowering plants. They are ovules with fertilized plant embryos inside of them.

India built a strong seed production infrastructure to meet the needs of the country. It is composed of various seed classes, such as Nucleus, Breeder, Foundation, and Certified seeds, each of which complies to certain quality standards at various stages to ensure the plentiful production of high-quality seeds for sustainable agriculture. High-quality seeds are the primary requirement for achieving grain production and they account for about 30% of the increase in yield (Sahu *et al.*, 2018).

16.10. Seed Multiplication System Of India

India's seed multiplication system is meticulously maintained to ensure the high quality and accessibility of a variety of seed types. The parent institute is in responsibility of providing the requested quantity of Breeder seed to the seed chain and guaranteeing the seed quality of newly released types. During this process, nucleus seed is created, which serves as the foundation for the creation of breeder seed and ensures the general quality of the country's seed supply. The Indian seed system typically operates on a three-generation basis (Breeder seed, Foundation seed, Certified seed), but in severe circumstances, where Foundation Seed Stage II or Certified Seed Stage II is developed, it can stretch to four or five generations.

In nucleus seed plots, one plant from each selected panicle is carefully nurtured in paired rows. Eight rows of the exact same variety (from bulk breeder seed) surround these plots. If any off-type plants are discovered, the entire pair of panicle progeny rows are destroyed. If an off-type plant with a different grain type is found (typically after flowering), the rows where the off-type plant is located and the neighboring rows on both sides are destroyed in order to prevent inadvertent cross-pollination with an off-type plant. A significant number of true-to-type panicles (at least 500) are chosen after careful selection based on their physical traits, homogeneity, and genetic purity in order to maintain the nucleus seed for the upcoming generation (Sahu, *et al.*,2018). The border row is especially made to restrict the flow of foreign pollen and is not taken into consideration for seed production. Panicle progeny rows are harvested, threshed, and meticulously inspected before being bulked as nucleus seed, which is used to make breeder seed.

Breeder seed (BS) is produced from nucleus seed, with skip rows inserted every 6–8 rows to aid in cultural operations and ensure proper roguing. Similar to nucleus seed plots, breeder seed plots include eight rows around them that are not taken into account for seed production when the crop is harvested. Simply get rid of any abnormal plants. The crop is constantly monitored by central and state monitoring teams, and breeder seed is required to always be 100% genetically pure. Breeder seed tags measure 12 by 6 cm and are golden in color. Foundation seed, often referred to as stage I foundation seed, is produced from breeder seeds. When the parent of the foundation seed is the same, it is referred to as foundation seed stage II. Both stage I and stage II of foundation seed must meet the same minimal seed criteria. Only when seed certifying bodies believe it is necessary to do so in order to address seed demand due to breeder seed shortages will stage II foundation seed be developed. State certification agencies monitor the genetic purity of foundation seed, which must be kept at 99.5%. For foundation seeds, white tags 15x7.5 cm in size are utilized. Certified seed is the progeny of Foundation Seed I or II. It may also be the offspring of certified seed as long as this reproduction doesn't extend more than three generations past stage I of foundation seed. Foundation seed is used to create stage I certified seed, and certified seed is used to create stage II certified seed. State certifying bodies scrutinize the 99% genetic purity requirement for certified seed.

For foundation seeds, white tags 15x7.5 cm in size are utilized. Certified seed is the progeny of Foundation Seed I or II. It may also be the offspring of certified seed as long as this reproduction doesn't extend more than three generations past stage I of foundation seed. Foundation seed is used to create stage I certified seed, and certified seed is used to create stage II certified seed. State certifying bodies scrutinize the 99% genetic purity requirement for certified seed. Seeds that have been approved are marked with blue tags that measure 15 x 7.5 cm. Additionally, India produces and sells Truthfully Labelled Seed (TFL Seed), a kind of seed that is not legally required to be certified but is nevertheless subject to minimal seed standards. Seed inspectors are crucial to quality control since they may assess TFL seed and halt sales if issues with its quality are discovered. The TFL Seed tag is opal green and is 15 x 10 cm.

16.11. Seed Multiplication Chain of India

The availability of high-quality seeds for various crop species is ensured in India by an organized process for seed multiplication. A detailed explanation of this chain is provided here, along with information on the organizations' duties, how they categorize the produced seed, and how they follow certification standards (Prasad et al., 2017).

Farmers are the first step in the process since they have specific criteria for crop varieties to meet their agricultural needs.

- ☐ **District Agriculture Officers (DAO):** District Agriculture Officers compile data on regional demand for certain seed varieties.

- ☐ **DDA (Deputy Director of Agriculture):** To determine the seed need at the district level, the DDA analyzes the varietal need and evaluates the overall state of agriculture.

- ☐ **Director of Agriculture:** The Director of Agriculture sets the precise quantity of seeds to be ordered at the state level based on local agricultural demands. Additionally, new varieties may be added to the seed chain by the director of agriculture.

❑ **DAC (Department of Agriculture and Cooperation):** At the national level, the Department of Agriculture and Cooperation coordinates the seed supply across the country and creates a national demand for the necessary number of seeds.

❑ **Company Charged With Producing Breeder Seeds:** Breeder seed businesses are responsible with producing the quantity of breeder seeds indicated in the contract. Breeder seeds are the original seeds that maintain the variety's genetic purity and superiority (Sahu *et al.*, 2018).

SSC & Director of Agriculture: It is the duty of the State Seed Corporation (SSC) and the Director of Agriculture to take the breeder seeds out of the breeder seed producer and add them to the seed chain for further multiplication and distribution.

16.11.1. Chain for Producing Seeds

❑ **District Agriculture Officers (DAOs)** coordinate the seed distribution, ensuring that the right varieties are made available to local farmers in accordance with their requirements.

❑ **Farmer (Growers):** The seeds are eventually given to the farmers, who then plant their crops using them.

16.11.2. Responsibilities of Seed Producing Organizations

Organizations engaged in the manufacturing of seeds must adhere to various requirements depending on the type of seed they are producing (Breeder seed, Foundation seed, Certified seed, or Truthfully Labelled seed). These commitments include maintaining genetic integrity, adhering to certification criteria, and guaranteeing the quality of the seeds they produce (Sahu *et al.*, 2018).

16.11.3. Seed Multiplication Generation System

The Seed Multiplication Generation System is essential to ensuring a steady and economical supply of seeds for agricultural output. It begins with the creation of a brand-new plant variety by a breeder, who at first has just a few nucleus seeds. This novel variety is made more widely available to farmers via a multi-step multiplication process inside of a controlled generation system (Nigam, *et al.*, 2004). For seed multiplication, there are several models, including:

❑ **A three-generation model:**

☆ **Breeder Seed:** These are the original seeds that the plant breeder produced.

☆ **Foundation Seed:** The breeder seed is grown at this point in the multiplication process to produce more seeds.

- ☆ **Certified Seed:** The final step is to produce certified seed, which is suitable for commercial distribution and use by farmers.
- ❏ **A model with four generations.**
    - ☆ **Breeder Seed:** The original seeds a breeder produced.
    - ☆ **The term "foundation seed" (I):** refers to the first round of multiplication, which produces a larger number of seeds.
    - ☆ **Foundation Seed (II):** An additional iteration of multiplication to increase the number of seeds.
    - ☆ **Certified Seed:** At this time, seeds have undergone testing and been determined to fulfill the standards for quantity and quality for commercial distribution.Breeder Seed: The first seeds that the breeder created.
- ❏ **The five-generational model:**
    - ☆ **Breeder Seeds:** These are the seeds that a plant breeder has developed as a base.
    - ☆ **Foundation Seed (I):** The first round of seed multiplication to increase seed output.
    - ☆ **Foundation Seed (II):** A second phase of reproduction for seed production.
    - ☆ Certified Seed (I) refers to the initial phase of certified seed production.
    - ☆ **Certified Seed (II):** This step makes sure there are enough high-quality seeds accessible for commercial use.

The Seed Multiplication Generation System is essential for scaling up the creation of new plant species. Depending on the model utilized, a number of steps may be necessary to transform breeder seeds into certified seeds suitable for widespread agricultural usage. By guaranteeing that the seeds maintain their high standards of quality and genetic purity throughout the process of multiplication, these models aid farmers and maintain agricultural production. (Nigam, *et al.*, 2004).

16.12. Seed Certification Process Phases

Six major phases make up the methodical and thorough process of seed certification:

16.12.1. Application Receipt and Review

- ❏ A seed producer or other organization seeking certification must submit an application as the first action in this phase.
- ❏ The application is rigorously evaluated to ensure that it is correct, thorough, and in compliance with all standards.

16.12.2. Seed Source, Class, and Compliance Verification

- ❏ During this stage, officials carefully review important factors such the seed's origin, classification (such as breeder, foundation, or certified seed), and adherence to requirements.
- ❏ The goal is to confirm that the seed crop being grown conforms with the stated standards.

16.12.3. Post-Harvest Supervision and Field Inspections

- ❏ The certification process requires field inspections. The objective is to guarantee that every step of the seed manufacturing process conforms with the necessary integrity and quality requirements.
- ❏ Inspectors analyze the standing crop in the fields by examining a variety of factors such plant health, genetic purity, and adherence to advised production procedures.

16.12.4. Vigorous Supervision

- ❏ After the harvest, strict supervision is maintained during post-harvest activities such as processing and packaging. In particular, strict supervision is required throughout the packaging and processing of seeds.
- ❏ The seed sample is an essential component of certification, and its quality and integrity must be maintained throughout these critical steps in order to prevent any compromise in the seeds' overall quality.

16.12.5. Seed Sample and Analysis

- ❏ Carefully selected samples are carefully gathered and studied in order to assess genetic purity and seed health, if necessary.
- ❏ Genetic purity tests demonstrate that the seeds are authentic to the variety that has been disclosed, and seed health tests guarantee that they meet the requirements for disease-freeness.

16.12.6. Certificate Issuance and use of Certification Tags, Bags, and Seals

- ❏ Following the successful completion of all earlier phases, the final step results in the issue of a certification certificate.
- ❏ To preserve the integrity and quality of the certified seeds, the seed bags are sealed, and certification tags are applied to show that the seeds meet the standards for quality. The six processes that make up the certification procedure for seeds are its core elements. They are essential in ensuring that seeds intended for use in agriculture adhere to stringent requirements for genetic quality, homogeneity, and purity. In the end, this rigorous certification process aids in the success and increased productivity of the agriculture sector (Nigam *et al.*, 2004).

16.13. Seed Certification Procedures

- ❏ **Source Verification:** All seed materials going into the seed production program are subject to source verification by the certification body. The agency may ask for specific paperwork when a seed producer submits an application or is subject to an initial seed crop inspection, including certification tags, seals, labels, seed containers, purchase records, and sales records. This supporting information shows that the seed used for planting comes from a source that has been approved by the agency and complies with the required seed class. Seed producers are required to maintain a source-verification record with the necessary data for agency staff to review.

- ❏ **Application for Certification:** After confirming their sources, seed producers must submit three copies of their "FORM-1" application for certification to the Assistant Director of Seed Certification in the relevant area. The FORM-1 is carried by divisional and zonal agency offices.

- ❏ **Timely Submission:** The FORM-1 must be submitted within 30 days of the sowing date or 15 days of the transplanting date, whichever is earlier.

- ❏ **The Payment of Certification Fees:** Seed producers must also file FORM-1 and pay any certification fees that could be necessary, such as those for registration, inspection, grow-out tests (if necessary), seed testing costs, etc.

- ❏ **Separate Applications for Each Variety:** A separate FORM-1 must be completed for the certification of each seed variety. A complete list of information should be included on the form, including the name and address of the seed producer, the season of production, the grower's name and address, the location of the seed plot, the crop or variety, the class of seeds used, the area under seed production, the specifics of the parental seed materials used with lot numbers, the date of sowing, and the details of the seed certification fees paid.

- ❏ **Maximum Certification Area:** A single application may include up to 25 acres of certification territory. Each additional area requires a separate application.

- ❏ **Non-Refundable costs:** Once the Seed Certification Officer has visited or investigated the seed plot, inspection and registration fees are non-refundable.

- ❏ **Coordination During Inspection:** Seed producers must coordinate in order for agency personnel to locate the seed plots during the initial inspection.

- ❏ Seed producers shall provide growers with information on agronomic practices, disease and pest management, and other essential aspects of seed production.

- ❏ **Field Standards Compliance:** At authorized seed processing facilities, only seed from plots that adhere to the relevant Field Standards for Certification is accepted.

- ❏ **Refraining from Admixture:** Admixture must be kept to a minimum during the harvest, threshing, and transportation of harvested seed products.

- ❏ **Timely Certification:** Seeds harvested from designated fields must be brought to the seed processing plant for timely certification within 2.5 months after the date of harvest.

- ❏ **Certification Requirements:** A seed lot can only be certified if it meets the necessary seed and field requirements.

These standards outline the specific steps involved in the seed certification process in order to ensure the quality and conformity of seeds intended for agricultural use (seednet.gov.in).

16.14. Constraints in Seed Production and Seed Research

Indian Seed Research and Production Challenges India's seed manufacturing and distribution system is the outcome of joint efforts from the governmental and commercial sectors. An elaborate institutional framework has been set up to ensure the production and distribution of high-quality seeds. Despite having a robust network of seed suppliers, it continues to be challenging to maintain a consistent flow of high-quality seeds from producers to farmers. Many states have Seed Replacement Rates (SRR) below 20% as a result of farmers continuing to use farm-saved seeds.

The Variety Replacement Rate (VRR), which is crucial for increasing agricultural output when utilizing superior seeds, is another significant component. Of the more than 900 high-yielding paddy hybrids and varieties that have been developed for commercial farming, only about 318 are currently actively involved in the seed production process (Sahu et al., 2018).

The effective production and distribution of seeds is hampered by a number of factors:

- ❏ **Buying Seeds from Untrustworthy Sources:** Farmers commonly buy seeds from vendors that might not be able to guarantee their quality, which could have unintended consequences on crop productivity.

- ❏ **Seed Quality Deterioration Over Time:** Seeds may become less effective in the field if they are duplicated over an extended period of time.

- ❏ **Limited Access to High-Quality Seeds:** Due to a lack of high-quality seeds, farmers are increasingly using saved seeds.

- ❏ **Low SRR or VRR:** The delayed uptake of new seed varieties limits the potential for increased agricultural output.

- ❑ **Lack of Technical Knowledge:** Many farmers are not sufficiently knowledgeable about modern seed production techniques.

- ❑ **Time-consuming Seed Quality Testing:** The time-consuming seed quality testing may make it more difficult for farmers to obtain seed.

- ❑ **Genetic Purity Uncertainty:** Farmers are concerned that some types developed with Marker-Assisted Selection (MAS) may not always ensure genetic purity.

References

Chand, Subhash, Kailash Chandra, and Champa Lal Khatik. "Varietal release, notification and denotification system in India." In *Plant Breeding-Current and Future Views*. IntechOpen, 2020.

Chauhan, J. S., Prasad, S. R., Pal, S., Choudhury, P. R., & Bhaskar, K. U. (2016). Seed production of field crops in India: Quality assurance, status, impact and way forward. *Indian Journal Agricultural Sciences*, 86(5), 563-79. https://seednet.gov.in/PDFFILES/Certification_Process.pdf

Kamble, U. R., Govind, P., Prasad Rajendra, S., Udayabhaskar, K., & Sripathy, K. V. (2015). Comparative studies of Indian seed laws with special reference to protection of plant varieties and farmers rights act.(In) Abstract book National seminar on harmonizing biodiversity and climate change: Challenges and opportunity. *ICAR-CIARI, Port Blair, A&N*, 17-19.

Nigam, S. N., Giri, D. Y., & Reddy, A. G. S. (2004). Groundnut seed production manual.

Prasad, S. R., Chauhan, J. S., & Sripathy, K. V. (2017, March). An overview of national and international seed quality assurance systems and strategies for energizing seed production chain of field crops in India. ICAR.

Reddy, C. R. (2007). *Seed system innovations in the semi-arid tropics of Andhra Pradesh*. ILRI (aka ILCA and ILRAD).

Sahu, R. K., Sah, R. P., Sanghamitra, P., Verma, R. L., Patil, N. K. B., Jena, M., ... & Singh, O. N. (2018). Quality seed production and maintenance breeding for enhancing rice yield. ICAR-NRRI.

Tonapi, V. A., Bhat, B. V., Kannababu, N., Elangovan, M., Umakanth, A. V., Kulkarni, R., ... & Rao, T. G. N. (2015). *Millet seed technology: seed production, quality control & legal compliance*. Hyderabad, India: Indian Institute of Millets Research.

Trivedi, R. K. (2012). Seed quality regulation and OECD varietal certification for export of seeds. In National Seed Congress on Welfare and Economic Prosperity of the Indian Farmers through Seeds. *Raipur, Chhattisgarh*, 21-23.

Chapter–17

Intellectual Property Rights in Plant Breeding and Bio-safety Issues in Plant Biotechnology

Thirumalraj S[1] , Tushar Arun Mohanty[1]*, Himakara Datta M[1] , Raiza Christina G[1] and Lavudya Sampath

Department of Genetics and Plant Breeding, TNAU, Coimbatore
**Corresponding author: tushararunmohanty@gmail.com*

17.1. Introduction

In the boundless world of agriculture and biotechnology, where science meets nature, there lies a delicate balance between innovation and preservation. The intersection of these two crucial facets is where the intriguing journey into the realm of Intellectual Property Rights (IPR) in plant breeding and the intricate web of biosafety issues in plant biotechnology unfolds. This chapter marks the beginning of our exploration into this captivating domain.

The evolution of agriculture, from its humble origins to the modern era, has witnessed remarkable transformations. Central to these transformations is the art and science of plant breeding, where humans have manipulated plant genomes to develop crops with enhanced traits, better yields, and improved resilience. In this era of rapid global population growth and environmental challenges, the significance of plant breeding and biotechnology has never been more apparent.

Plant biotechnology represents a powerful tool in agriculture, offering the potential to enhance crop yields, resist pests, and adapt to adverse environmental conditions. These advancements, including genetically modified organisms

(GMOs), hold great promise. However, as we harness this technology, we must navigate critical ethical and biosafety considerations. Intellectual Property Rights (IPR) mechanisms are essential for protecting the innovations of breeders and biotechnologists, driving research and development. Still, they also raise concerns about equitable access and benefit-sharing, particularly for developing nations. Biosafety is paramount, necessitating rigorous risk assessments to safeguard human health, biodiversity, and the environment. Coexisting GMO crops with conventional and organic farming adds complexity to the agricultural landscape. In this evolving field, the delicate balance between innovation, IPR, biosafety, and ethical responsibility is crucial to ensure progress aligns with environmental and societal well-being.

In this chapter, we embark on a comprehensive journey into the world of IPR in plant breeding and biosafety in plant biotechnology. We delve into historical context, key players, legal frameworks, and strategies. Ethical dimensions, societal implications, and global impacts are also explored.

As we turn the pages ahead, we invite you to join us in uncovering the complexities, opportunities, and challenges that shape the tapestry of intellectual property rights in plant breeding and biosafety issues in plant biotechnology. Together, we embark on a journey seeking to harmonize innovation with conservation, progress with ethics, and science with stewardship in the pursuit of sustainable agriculture for future generations.

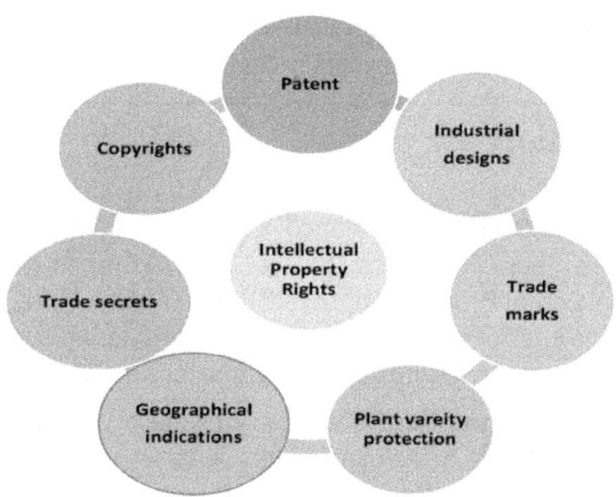

Figure 17.1: Represents different types of intellectual property rights.

17.2 Intellectual Property Rights (IPR) in Plant Breeding

Intellectual Property Rights (IPR) in plant breeding encompasses legal protections and mechanisms that grant breeders and developers of new plant varieties exclusive rights to their innovations. These rights are designed to encourage investment in research and development, fostering innovation in agriculture. Key aspects of IPR in plant breeding include:

17.2.1. Plant Variety Protection (PVP): PVP is a form of IPR that grants breeders exclusive rights to a newly developed plant variety. To receive PVP, the variety must be distinct, uniform, stable, and novel. This protection typically lasts for 20-25 years, during which others are prohibited from reproducing, selling, or using the protected variety without permission. PVP encourages breeders to invest in the development of new, improved plant varieties.

17.2.2. Plant Patents: Plant patents are another form of IPR that focus on the unique qualities of a plant. They are granted by the government for new, distinct, and asexually reproduced plant varieties. Breeders with plant patents have exclusive rights to propagate, sell, and use the patented plant. Plant patents are typically valid for 20 years from the date of filing.

17.2.3. Utility Patents: Utility patents cover innovations related to plant breeding processes, methods, and genetic sequences. These patents protect the methods used to develop new plant varieties or specific genetic traits. Utility patents are valuable in cases where the innovation is not limited to a specific plant variety but extends to broader applications in plant breeding and biotechnology.

17.2.4. Trade Secrets: Some breeders opt to keep their breeding methods, genetic information, or other proprietary knowledge as trade secrets. They rely on non-disclosure agreements and strict confidentiality measures to protect their innovations. Trade secrets provide ongoing protection without a defined expiration date, as long as the information remains confidential.

17.2.5. Licensing Agreements: Breeders often enter into licensing agreements to grant others the right to use, develop, or commercialize their plant varieties. These agreements may involve royalties or other considerations. Licensing can facilitate the dissemination of new varieties and technologies while allowing breeders to generate revenue.

17.2.6. Access and Benefit-Sharing (ABS): ABS mechanisms are important when plant breeding involves genetic resources sourced from indigenous or local communities. These arrangements ensure that those providing genetic materials are fairly compensated for their contributions. ABS agreements aim to strike a balance between benefiting breeders and respecting the rights and knowledge of local communities.

IPR in plant breeding serves as a vital incentive for breeders to invest time and resources in developing improved plant varieties. However, it also sparks debates about accessibility and equity, especially in the context of food security and sustainable agriculture. Balancing innovation with the broader public interest remains an ongoing challenge in the field of plant breeding and IPR.

17.3. Impact and Limitations of IPR Regimes in Plant Breeding

Limited experience in developing nations provides insight into the effects and constraints of Plant Variety Protection (PVP). Although PVP offers advantages, its efficacy varies depending on the specific circumstances.

- ❐ **Protection for Small Enterprises:** In countries where domestic private seed industries develop new varieties of non-hybrid crops, PVP ensures that small companies' varieties remain inaccessible to competitors. For instance, it aids in safeguarding rice varieties in Colombia and wheat varieties in Argentina.

- ❐ **Security for Hybrid Varieties:** In nations like India and China, where securing inbred lines physically is challenging due to close competition among enterprises, PVP is valuable for preserving hybrid varieties.

- ❐ **Control over Informal Seed Sales:** In regions dominated by larger farmers, PVP can help regulate large-scale informal seed sales of protected varieties and potentially impose restrictions on seed saving.

- ❐ **Support for Public Research Institutes:** A well-functioning PVP system equips public research institutes with tools to guide the deployment of their varieties and encourages foreign companies to collaborate by sharing germplasm with local partners.

- ❐ **Additional Security for Export Crops:** PVP can provide extra security for specialty export crops under contract. However, it's essential to recognize that merely establishing a PVP framework isn't enough to stimulate significant investments in plant breeding in developing countries.

In countries like India, the significant emphasis on "farmer's rights" in the PVP Act limits profits from developing and selling pure-line varieties. As a result, private sector plant breeders may concentrate on the highly profitable hybrid seed sector, leaving the development of pure-line varieties for high-input and marginal areas to the public sector and farmers. Decisions regarding PVP should be made in conjunction with broader policy support to ensure a diverse and competitive private seed sector and a responsive public agricultural research system.

Key factors include a well-functioning legal system that enables enforceable contracts with seed growers and merchants, the promotion of responsible business practices, and support for professional associations in agribusiness. Effective seed regulations are also crucial, including mechanisms for variety release and approval, seed quality control, and consumer education. These regulations should empower seed producers to take on increasing responsibility.

17.4.The Nature of IPRS for Plant Breeding

❏ **Challenges in Plant Variety Protection:** Safeguarding plant varieties using traditional patent systems poses difficulties due to issues such as expressing novelty, establishing inventive steps, and providing accurate descriptions for replication by those skilled in the field.

❏ **Sui Generis Protection:** Given the distinctive nature of plant breeding, many countries have implemented sui generis systems specifically designed for this purpose. The primary focus of these systems is to protect plant varieties.

❏ **UPOV Conventions:** The passage mentions the UPOV Conventions, which were established to offer effective protection for plant varieties and promote the development of new plant varieties for the benefit of society. It discusses two key UPOV Conventions: one from 1978 and another from 1991, each with its own set of requirements and provisions.

❏ **Fundamental Requirements for Protection:** Under UPOV, a plant variety can receive protection if it meets the criteria of being distinct, uniform, and stable (DUS) and has not been previously marketed without the breeder's consent. For utility patents, the variety must be new, non-obvious, and useful for industrial applications (including agriculture). Usually, a sample of the variety is deposited with the patent office to fulfill the disclosure requirement.

❏ **Control over Usage:** UPOV Conventions provide a mechanism to prevent unauthorized commercial production of protected plant varieties. UPOV 1991 grants plant breeders the authority to manage the utilization of harvested materials if they have not received royalties on the propagating materials.

❏ **Competing Breeding Programs:** In the context of UPOV 1978, protected varieties can be freely employed in competitors' breeding programs. However, UPOV 1991 addresses situations where new varieties exhibit slight variations and allows the original variety's owner to share in the benefits. In contrast, utility patents do not permit the use of protected varieties in other breeding programs without permission.

❏ **Seed saving and Exchange by Farmers:** The three options discussed (UPOV 1978, UPOV 1991, and utility patents) have distinct rules regarding farmers' ability to save and exchange seeds. UPOV 1978 only deals with the commercial sale of seeds and does not impose restrictions on seed saving or non-commercial seed exchanges among farmers. UPOV 1991 prohibits the multiplication of protected varieties but allows specific exceptions for farmers to use such varieties on their own lands. Seed varieties protected by utility patents cannot be saved or exchanged.

- ❏ **IPR Systems in Developing Countries:** Developing countries can choose to adopt systems resembling UPOV or create their own sui generis systems tailored to their specific needs. Some countries are contemplating legislative changes to align with UPOV 1991.

- ❏ **Regional Approaches:** Certain regions, such as the African Intellectual Property Organization (OAPI), have implemented regional PVP (Plant Variety Protection) systems that adhere to UPOV principles. These systems simplify the process of protecting plant varieties across multiple countries.

17.5 Opportunities for Development Agencies

- ❏ **Customized Protection Systems:** Development agencies can have a pivotal role in aiding policymakers in crafting intellectual property rights (IPR) protection systems that align with the unique requirements and capacities of a country's seed systems. These systems should take into account variations among export crops, domestic commercial commodities, and subsistence crops.

- ❏ **Beyond Regulatory Structures:** Although establishing a regulatory framework is vital, it constitutes just one component in supporting commercial breeding and seed supply. Additional efforts are imperative to cultivate the administrative, technical, and human resources necessary for the effective implementation of these systems. Active engagement of all stakeholders is essential.

- ❏ **Impact on Research Focus:** The integration of revenue generation through IPRs into public plant breeding can reshape the allocation of funds and breeding strategies. There's a potential risk that resources might be redirected away from critical disciplines like soil science, social sciences, and plant pathology toward breeding endeavors. Prioritization may lean toward crops with high seed production value, potentially sidelining crops crucial for ensuring nutritional security within the population.

- ❏ **Shift in Research Priorities:** The pursuit of revenue generation may induce a shift in research priorities, favoring commercial farmers and hybrid varieties over resource-poor farmers and open-pollinated varieties. While this realignment may align with changes in national agricultural policy, it may not necessarily support equity and broader agricultural production objectives.

- ❏ **Financial Considerations:** The implementation of IPRs in plant breeding comes with substantial costs, encompassing application and maintenance fees. Safeguarding IPRs, especially when contending with powerful commercial entities, can also prove financially burdensome. Governments may need to allocate funding for research if National

Agricultural Research Institutes (NARIs) are unable to protect their innovations.

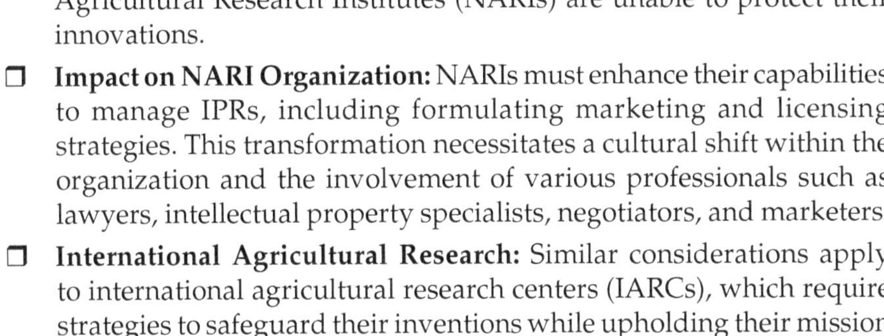

- **Impact on NARI Organization:** NARIs must enhance their capabilities to manage IPRs, including formulating marketing and licensing strategies. This transformation necessitates a cultural shift within the organization and the involvement of various professionals such as lawyers, intellectual property specialists, negotiators, and marketers.

- **International Agricultural Research:** Similar considerations apply to international agricultural research centers (IARCs), which require strategies to safeguard their inventions while upholding their mission of poverty alleviation. Ensuring access to protected technologies must not hinder reaching resource-poor populations. Collaborations and agreements must address these challenges.

- **Flexibility in TRIPS Agreement:** The TRIPS Agreement affords flexibility for countries to tailor their IPR regimes to their specific circumstances, particularly in agriculture. Nonetheless, there is pressure on developing nations to adopt more rigid IPR regimes in bilateral trade negotiations and discussions within organizations like WIPO.

- **Justification for Enhanced IPRs:** Policymakers should meticulously evaluate the role of IPRs in agricultural development and assess their impact on incentives for both domestic and foreign investments. Any strengthening of IPRs should be grounded in a thorough assessment of the national agricultural and breeding sectors and should involve consultations with key stakeholders.

17.6 IP Protection for Plant Varieties

- **Plant Breeders' Rights (PBR):** PBR, also referred to as Plant Variety Protection (PVP), constitutes a specialized framework designed to ensure the uniqueness of plant and mushroom varieties. Administered under the International Convention for the Protection of New Varieties of Plants (UPOV), it provides breeders with exclusive rights when their variety exhibits clear, consistent, and stable characteristics (D.U.S.). Typically, these rights endure for a period of 20 to 25 years, granting breeders the authority to grant licenses for multiplication and agricultural utilization.

- **Patents:** Patents serve as a mechanism for safeguarding novel inventions characterized by their innovative nature. They confer control over the utilization of the patented invention. It is noteworthy that patent laws and PBR function independently. While certain regions impose restrictions on patenting plant varieties and "essentially biological processes," others allow patents for technologies such as transgenic constructs and gene editing.

- ❏ **Brand Names and Trademarks:** Associating plant varieties with brand names or trademarks empowers breeders to oversee the conditions under which the variety is marketed under that specific name. For example, the trademark "Pink Lady" is linked with the Cripps Pink apple variety. Unlike PBR, brand names and trademarks do not come with predefined expiration periods but necessitate distinct protective measures.

- ❏ **Private Law Contracts:** Legal agreements established under private law can be utilized to enforce compliance with Good Agricultural Practices (GAP) and other conditions related to plant varieties. These contracts may encompass obligations governing the utilization of the variety and provide the owner with the prerogative to inspect the fields of farmers. Contractual cultivation represents another approach, wherein farmers are contracted to cultivate or propagate the variety based on specific arrangements.

17.7 Intellectual Property Rights (IPRs) and Traditional Knowledge

The nature of Intellectual Property Rights (IPRs) and their impact on traditional knowledge often give rise to concerns regarding fairness and potential exploitation within indigenous communities. These communities possess knowledge and innovations that are essential for the sustainable conservation of biodiversity. Two critical aspects require careful consideration:

Firstly, IPRs inherently have elements that can lead to injustices against holders of traditional knowledge. Traditional knowledge is distinctive in its continuous evolution, resulting in ongoing improvements and practical solutions developed for survival. It often does not adhere to conventional Western scientific methodologies. Additionally, its value is deeply intertwined with religious, moral, cultural, political, and commercial aspects, closely linked to the habitat and environment. In many cases, traditional knowledge lacks tangible material embodiment and tends to yield informal products. These unique characteristics make it challenging to protect using existing IPR frameworks, which often do not comprehensively address all aspects of traditional knowledge. Therefore, safeguarding traditional knowledge is best achieved through a dedicated sui generis system.

Secondly, to address the existing imbalance in trade-related intellectual property rights (TRIPS) and to protect various innovation systems, consideration could be given to amending Article 27.3(b) of the TRIPS Agreement. Such an amendment would require members of the World Trade Organization (WTO) to establish effective sui generis systems for the protection of traditional knowledge and folklore. This protection should be designed in alignment with international agreements like the Convention on Biological Diversity (CBD), the International Treaty on Plant Genetic Resources for Food and Agriculture (ITPGRFA), and existing regional and national regulatory frameworks. This approach would take into account the rights and interests of indigenous and local communities that have nurtured traditional knowledge, ensuring its preservation and fair treatment within the realm of intellectual property rights.

17.8 Agricultural Biotechnology

The origins of modern agricultural biotechnology can be traced back to 1983 when the first successful transfer of a plant gene from one species to another occurred. Notably, one of the early breakthroughs in biotechnology that garnered significant attention involved genetically modified bacteria. This development, exemplified by Chakrabarty's patent application in 1981, was designed to facilitate the rapid cleanup of oil spills, highlighting the direct relevance of biotechnology to environmental conservation. As a result, both agricultural and environmental applications have shown great promise in delivering benefits to ecosystems. Biotechnology is anticipated to contribute to improved nutrition and offer a safer alternative to many of the currently employed pesticides and herbicides (Pray et al., 2005).

In the private sector, agricultural biotechnology primarily emphasizes technologies related to crop protection and environmental sustainability, rather than techniques for enhancing crop yields. It's important to understand this distinction in a relative sense, as the resources required for fundamental research, germplasm, and associated costs for developing varieties with improved yields far exceed those needed for creating disease and pest-resistant varieties (Bent, 1987). However, it's noteworthy that this sector has experienced rapid growth in recent years, partly due to government involvement. Advanced agricultural biotechnology mainly unfolds in industrialized nations with more extensive research and development capabilities. The ownership and utilization of Intellectual Property Rights (IPRs) play a crucial role in determining the success of any technological innovation introduced to the market, serving as a driving force behind sustained technological advancement and enhancing a nation's industrial competitiveness.

17.9 Biodiversity

India holds substantial commercial potential in agricultural biodiversity, offering sustainable avenues for socioeconomic progress. The commercialization of valuable plants and animals presents a viable strategy for poverty alleviation in the country. Despite contributing around 10% of the world's agricultural production, India's share in global agricultural trade remains below 1%, primarily due to protective agricultural policies (Sharma *et al.*, 2003).

In the realm of biotechnology, certain developed nations like the United States and Australia extend patent protection to microbial processes and plant varieties. However, patenting new life forms has proven to be a complex endeavor, with significant disparities among countries. The trend toward patenting living organisms has ignited vigorous debates on regional, national, and international levels.

Opponents of life form patents contend that using the patent system to reward scientific work in biological fields is inappropriate, as living organisms differ fundamentally from inanimate materials. Moreover, provisions are necessary to ensure prior consent and equitable benefit-sharing with indigenous and local communities that have traditionally safeguarded these resources. The potential adverse consequences of granting private patent rights over genetic resources raise concerns in many biodiverse countries, prompting the need for sui generis systems or a combination thereof.

The most prevalent sui generis system for protecting plant varieties is the International Convention for the Protection of New Varieties of Plants (UPOV Convention). Despite evolving independently of patent protection, plant variety protection is considered more suitable for the unique nature and characteristics of agricultural innovations. Nevertheless, heightened levels of protection have raised concerns similar to those observed in the patent realm. Revisions to the UPOV Convention have generally aimed to progressively strengthen plant breeders' rights (Ravishankar and Archak, 1999; Alston and Venner, 2002).

The conservation and sustainable development of technologies encompass various methods, including tissue culture, field-based propagation, protoplast fusion, and cryopreservation. Common mechanisms for transferring technologies include collaborative research and development (R&D), training nationals at foreign universities and institutions, and establishing technology partnerships through biodiversity-prospecting agreements.

The exception to patentability outlined in Article 27.3(b) also provides opportunities for sui generis protection of plant varieties. This provision mandates that members must offer protection for plant varieties, either through patents or an "effective sui generis system." The interpretation and application of these provisions regarding plant variety protection hold significant implications for the implementation of the Convention on Biological Diversity (CBD). Rights to information, as stipulated in the TRIPS Agreement, impact the sharing of benefits arising from the utilization of genetic resources. For instance, although a substantial portion of in-situ biodiversity and associated traditional knowledge, innovations, and practices are found in developing countries, most patents related to biological resources are granted for research conducted in developed nations. A well-defined sui generis protection system may serve as a tool to advance the CBD's objectives, including facilitating access and benefit-sharing and promoting technology transfer.

17.10 The International Convention for the Protection of new Varieties of Plants (UPOV) Convention

The International Convention for the Protection of New Varieties of Plants (UPOV Convention) traces its origins back to Paris in 1961 and came into force in 1968. Over time, it underwent revisions in Geneva in 1972, 1978, and 1991. The 1978 Act became effective in 1981, while the 1991 Act was enforced in April 1998. Currently, UPOV includes 38 member states, with 29 adhering to the 1978 Act and eight to the 1991 Act.

UPOV serves as the framework for safeguarding intellectual property rights related to plant varieties, often referred to as plant variety rights or plant breeders' rights (PBRs). To qualify for protection, a plant variety must meet specific criteria. It should display distinct, stable, and uniform characteristics (UPOV, 1991), or homogeneity regarding the specific feature of its sexual reproduction or vegetative propagation (UPOV, 1978). Additionally, it must be novel, meaning it has not been available for sale or marketed with the breeder's consent or without infringing on the breeder's rights, either in the source country or for an extended period in any other country (Wright and Parley, 2006; Das, 2011).

UPOV (1978) defines the scope of protection as the breeder's right to prior authorization for specific acts, including production for commercial marketing, offering for sale, and marketing of the reproductive or vegetative propagating material of the variety (Article 5). The UPOV (1991) version expands the scope of breeders' rights in two significant ways. Firstly, it broadens the range of acts requiring prior authorization from the breeder, encompassing production or reproduction, conditioning for propagation, offering for sale, selling, marketing, exporting, importing, and stocking for these purposes (Article 14). Secondly, these acts pertain not only to reproductive or vegetative propagating material, as in the 1978 version, but also cover harvested material obtained through the use of propagating material and so-called "essentially derived" varieties (Dutfield, 2002; Das, 2011).

The International Convention for the Protection of New Varieties of Plants (UPOV Convention) holds significance as it establishes a legal framework for safeguarding plant varieties developed by commercial plant breeders. This framework introduces "plant breeders' rights," a unique form of intellectual property rights that incentivizes the seed industry, similar to patents, without granting complete monopolies (Cullet and Raja, 2004).

However, it's important to note that knowledge related to biological processes and biological material doesn't qualify as inventions. Under the TRIPS Agreement, member countries can exclude plants and animals, as well as essentially biological processes for creating new plants and animals, from patentability. Nevertheless, TRIPS Agreement member countries are obligated to provide some form of protection for plant varieties, whether through patents or an effective sui generis system, or a combination of both.

Enforcing intellectual property rights goes beyond protection levels and scope. It's crucial for regulating and facilitating trade in genetically modified (GM) crop varieties. Additionally, the increasing "privatization of science" poses new management challenges, especially for research institutions in developing countries lacking the resources to effectively manage proprietary knowledge. Negotiating skills, administrative capabilities, and bureaucratic limitations often serve as tangible barriers to acquiring, negotiating, and protecting intellectual property rights, hindering access to certain strategic technologies.

Intellectual property policies are also closely linked to broader economic policies, such as creating an environment conducive to foreign direct investment and facilitating the participation of foreign firms in domestic markets. Technology transfer is another interconnected aspect of intellectual property rights, playing a vital role in transferring valuable knowledge with direct economic value from research and development institutes to industry. Genomics-centric biology, driven by hybrid varieties, reshapes the balance between biomedical researchers' interests, private sector market involvement, and the public good, as exemplified by the discovery of the Human Genome Sequence, Incyte, and Sequena (Ramasami, 2009).

17.11 Scope of Patentability

The extent of what can be patented has a significant impact on protecting investments and ensuring access to inventions by others. Many developing countries utilize exceptions and face challenges when enforcing existing patents, which can lead foreign investors to question the adequacy of property rights protection for genetically modified (GM) technologies. On the other hand, extensive patent protection can raise concerns related to food security, biosafety, biodiversity conservation, and socioeconomic issues. A situation known as "patent thickets" arises when numerous patents cover various innovations in GM crops, especially when a single innovation is subject to multiple patents, potentially impeding further research (Yamin, 2003). Therefore, while patent protection in agricultural research is not a new concern, proprietary claims are expanding, particularly concerning research tools. Developing countries often focus their research and development efforts on minor innovations and enhancements in existing technologies, and these efforts may face obstacles due to robust patent protection (Rangasamy and Elumalai, 2009).

Securing patents can be a costly endeavor, and patents are typically sought in countries where significant returns are anticipated from commercializing the patented subject matter. The country where competitors manufacture or reside is an additional factor in deciding where to file patents. Key biotechnology patents are seldom filed in developing countries, except in cases where major crops like soybean, canola, and cotton are extensively cultivated, as observed in Argentina, Brazil, and China (Mayer, 2003).

Furthermore, heightened patent protection can pose challenges in achieving the objectives of other international agreements, particularly the 1992 Convention on Biodiversity (CBD). The CBD recognizes states' sovereign rights over their natural resources, including genetic resources, and stipulates that access to such resources must be based on prior informed consent and mutually agreed terms. It also emphasizes the need for prior consent and equitable benefit-sharing with indigenous and local communities that have historically safeguarded these resources. The potential adverse consequences of patents, which grant private rights over genetic resources, raise concerns in numerous biodiversity-rich countries (CBD, 1999; International Centre for Trade and Sustainable Development).

17.12 Biosafety Issues in Plant Biotechnology:

Biosafety issues in plant biotechnology pertain to the safety and potential risks associated with genetically modified (GM) crops and their release into the environment, as well as their use in agriculture. These issues are critical to ensure that the introduction of GM crops does not harm human health, biodiversity, or the environment. Some key biosafety concerns in plant biotechnology include:

- ❑ **Environmental Impact:** One of the primary concerns is the potential environmental impact of GM crops. These crops can potentially crossbreed with wild relatives or other non-GM crops, leading to the spread of modified genes in natural ecosystems. This gene flow can disrupt local biodiversity and ecological balances.

- ❑ **Pest Resistance:** GM crops engineered to resist pests often produce insecticidal proteins. Over time, repeated exposure to these proteins can lead to the development of pest resistance. This not only reduces the effectiveness of GM crops but can also result in the evolution of super pests that are more challenging to control.

- ❑ **Weed Resistance:** Similarly, GM crops engineered for herbicide tolerance can lead to the emergence of herbicide-resistant weeds. This can necessitate the use of stronger herbicides, which can have detrimental effects on the environment and human health.

- ❑ **Gene Flow Management:** Controlling the flow of GM genes to non-GM crops or wild relatives is a significant challenge. Pollen drift, seed dispersal, and other factors can facilitate gene flow. Effective strategies are required to minimize this unintended spread.

- ❑ **Human Health and Allergenicity:** Ensuring the safety of GM crops for human consumption is a critical biosafety consideration. This involves assessing potential allergenicity, toxicity, and unintended health effects associated with consuming GM products.

- ❑ **Food Safety:** Beyond human health concerns, biosafety evaluations also encompass the safety of GM crops as food products. Rigorous testing and labeling are essential to inform consumers about the presence of GM ingredients and to prevent unintended consumption by individuals with allergies or dietary restrictions.

❏ **Regulatory Frameworks:** Establishing comprehensive and transparent regulatory frameworks for the approval, monitoring, and labeling of GM crops is essential. These frameworks must adapt to evolving technologies and scientific knowledge while ensuring the protection of public health and the environment.

❏ **Ethical and Societal Concerns:** Biosafety discussions often extend into ethical considerations. This includes assessing the impact of GM crops on small-scale farmers, indigenous communities, and traditional agricultural practices. Additionally, addressing societal perceptions of GM crops, their benefits, and potential risks is crucial.

❏ **Long-Term Environmental Effects:** Assessing the potential long-term environmental effects of GM crop cultivation, such as impacts on soil health, water quality, and ecosystem dynamics, requires ongoing research and monitoring.

❏ **Coexistence with Non-GM Crops:** The coexistence of GM and non-GM crops, including organic farming, is a complex issue. Ensuring that GM crops do not contaminate non-GM crops and that labeling is accurate requires careful planning and management.

❏ **Regulatory Compliance:** Enforcing compliance with biosafety regulations, monitoring GM crop field trials, and conducting post-release assessments are vital to minimizing risks and ensuring that GM crops meet safety standards.

Addressing these complex biosafety issues involves a multidisciplinary approach. It requires rigorous scientific research, risk assessment, regulatory oversight, public engagement, and responsible management practices to balance the potential benefits of plant biotechnology with the imperative of safeguarding the environment, human health, and societal well-being.

17.13 Issues Due to Antibiotic Resistance Genes

Bio-safety concerns related to antibiotic resistance genes (ARGs) in plant biotechnology revolve around the risk of ARG transfer to environmental bacteria, the potential persistence of these genes, and their unintended spread. Some genetically modified (GM) crops use ARGs as selectable markers during the modification process, which could lead to the development of antibiotic-resistant bacteria in the ecosystem if these genes are transferred. While the likelihood of ARG transfer to human pathogens is considered low, it remains a concern for human health. Regulatory oversight is crucial, with many countries having established guidelines for the use of ARGs in GM crops. To mitigate risks, researchers are actively developing alternative selectable markers, exploring gene editing technologies, and encouraging the phasing out of ARGs to ensure the responsible and safe use of biotechnology in agriculture.

17.14 Food Safety on Gm Crops

Food safety is a paramount concern in the context of genetically modified (GM) crops. GM crops are engineered to possess specific traits, such as resistance to pests, tolerance to herbicides, or improved nutritional content. Ensuring the safety of these crops for human consumption is essential. Before GM crops are allowed on the market, they undergo extensive safety assessments and regulatory reviews by government agencies. These evaluations consider allergenicity, toxicity, nutritional quality, and potential environmental impacts. GM crops developed for improved nutritional content are rigorously tested to ensure that the changes in nutrient composition are beneficial and do not compromise overall nutritional quality. Ongoing monitoring and surveillance are crucial to detect any adverse effects that may not have been apparent during pre-market assessments. International organizations provide guidelines and standards for the safety assessment of GM foods, ensuring consistency in regulatory approaches worldwide. Ultimately, while GM crops offer potential benefits for food security and nutrition, their safety is rigorously evaluated to protect human health and the environment.

17.15 Issues In Horizontal Gene Transfer

Bio-safety concerns surrounding horizontal gene transfer (HGT) in plant biotechnology are significant, primarily due to the potential unintended consequences and ecological impacts. HGT raises the specter of GM traits spreading to non-GM plants or related species, potentially altering natural ecosystems and biodiversity. The transferred genes could make recipient plants more invasive or competitive, raising concerns about ecological disruptions. Moreover, multiple GM traits may be transferred simultaneously, exacerbating their ecological influence. The transfer of GM traits to non-target organisms and altered plant-microbe interactions in the soil further complicates the situation. Effective containment of GM plants to prevent HGT is challenging, and regulatory oversight is essential. To address these concerns, researchers and regulators conduct thorough risk assessments, implement containment measures, and monitor the environmental impact of GM crops. Ongoing research enhances our understanding of HGT mechanisms, enabling more informed and responsible decision-making in plant biotechnology.

17.16 Issues in Super Weeds Due Toherbicide Resistance Genes

Bio-safety concerns surrounding superweeds, which have developed resistance to herbicides, are substantial due to the unintended ecological and agricultural consequences of herbicide-resistant genes. The emergence of these superweeds often leads to increased herbicide usage, potentially resulting in chemical runoff, soil contamination, and harm to non-target organisms.

Furthermore, the evolution of herbicide-resistant weed populations due to the presence of resistant genes in crops can disrupt farming practices, reduce crop yields, and pose economic challenges for farmers. The gene flow of herbicide resistance to wild plant relatives has ecological consequences, as it can disrupt local ecosystems and alter plant communities. Herbicide drift and the unintentional transfer of herbicide-resistant genes to non-target plants are additional concerns. Sustainable agricultural practices and integrated weed management strategies are crucial to mitigate these bio-safety issues, emphasizing the need for diversified cropping systems and non-chemical weed control methods. Additionally, regulatory measures and industry collaboration are essential to monitor and manage herbicide resistance effectively and promote responsible herbicide use in agriculture.

17.17 Intellectual Property Rights (IPR) for Transgenic Varieties: Implications and Challenges

The discourse surrounding IPRs related to transgenic varieties primarily revolves around patents for the genes, tools, and processes employed in genetic transformation. While this paper does not delve deeply into these intricate issues, it explores the ramifications of introducing Plant Variety Protection (PVP) legislation in developing countries concerning the level of security required to foster the development of commercially viable transgenic varieties for resource-limited farmers.

17.17.1. Positive Examples

Insect-Resistant Bt Cotton and Herbicide-Tolerant Soybeans: Instances of insect-resistant Bt cotton and herbicide-tolerant Roundup Ready (RR) soybeans provide promising evidence of the potential benefits of IPRs for transgenic varieties.

17.17.2. Challenges in India

Bt Cotton in India: In India, Bt cotton varieties were developed through a partnership between Monsanto and India's largest seed company, Mahyco. These varieties were introduced before the enactment of the PVP law, and the absence of gene patenting in India presented challenges. Reports of underground seed markets and clandestine breeding efforts have been widespread. Enforcement capacities have yet to be tested, and farmers still have the liberty to save and exchange seeds.

17.17.3. China's Complex Situation

Bt Cotton in China: China's experience with Bt cotton is intricate. While Bt cotton varieties are accessible through various collaborations, many face competition from unlawful seed multiplication and marketing. Initially, China excluded cotton from its PVP law, and there is limited enforcement capability.

Protection Hurdles: Monsanto has hesitated to introduce its latest Bt technology in China due to protection challenges. The company appears to be shifting toward licensing technology to Chinese companies.

Access Management: In smaller cotton markets, access to Bt technology can be managed by controlling the output system. In certain countries such as Colombia and South Africa, stringent controls have been imposed through grower registration and designated ginneries.

Issues with Herbicide-Tolerant Soybeans: Herbicide-tolerant soybeans, a widely cultivated transgenic crop, encounter challenges in South America. Despite gene patents, illegal seed saving and trading are prevalent. Monsanto has initiated legal actions for patent infringement against shipments of soybeans to European countries where the technology is patented.

Royalty Debate: One debated solution is the imposition of royalties on farmers based on grain sales at harvest, with collection managed by various stakeholders. A key point of contention revolves around farmers asserting their legal right to save seeds.

Managing Access to Transgenes: Managing access to transgenes by rival seed firms generally poses no significant issues. A combination of bio-safety regulations, field trials, and variety registration requirements curbs misappropriation.

Enforcement Challenges: Safeguarding transgenic varieties from illicit seed producers presents a more significant challenge, as observed in India and China. Enforcement of basic seed laws, variety approval procedures, and seed dealer licensing would help limit this issue.

17.18 Conclusion

In conclusion, the intersection of intellectual property rights (IPR) in plant breeding and bio-safety issues in plant biotechnology is a multifaceted and critical aspect of modern agriculture and innovation. IPR, including patents and plant variety protection, serves as a fundamental incentive for plant breeders and biotechnologists, encouraging investment in research and development of new crop varieties and biotechnological advancements. However, achieving a delicate balance between protecting innovators' rights and ensuring equitable access to genetic resources and benefits, especially for indigenous communities and biodiverse regions, is paramount. International agreements, such as the Convention on Biological Diversity (CBD), underscore the importance of obtaining informed consent and fair benefit-sharing.

Moreover, as genetically modified (GM) crops become more prevalent, robust bio-safety regulations are essential to evaluate and manage potential risks, safeguarding ecosystems, non-target species, and human health. Simultaneously, preserving food security and promoting diverse and resilient food supplies remain critical goals. Collaboration, technology transfer, ethical considerations, and comprehensive education are key to addressing these complex challenges effectively.

As biotechnology continues to advance, IPR systems must adapt to encompass emerging technologies like gene editing, while also addressing issues related to data ownership and sharing. In navigating this intricate landscape, it is imperative for governments, international organizations, scientists, and the private sector to work collaboratively in developing balanced and responsive regulatory frameworks. These frameworks should foster innovation while upholding environmental, ethical, and social values, ultimately prioritizing equitable access, benefit-sharing, and long-term sustainability to address the pressing global challenges of food security and biodiversity conservation.

References

Aerts, R.J. (2018). The European Commission's notice on Directive 98/44 and the European Patent Organization's response: The unpredictable interaction of EU and EPC law. Journal of Intellectual Property Law and Practice, 13, 800–805.

Bonny, S. (2016). Genetically Modified Herbicide-Tolerant Crops, Weeds, and Herbicides: Overview and Impact. Environmental Management, 57, 31–48.

Carrière, Y., Brown, Z.S., Downes, S.J., Gujar, G., Epstein, G., Omoto, C., ... & Carroll, S.P. (2020). Governing evolution: A socioecological comparison of resistance management for insecticidal transgenic Bt crops among four countries. Ambio, 49, 1–16.

Carrière, Y., Degain, B.A., Harpold, V.S., Unnithan, G.C., Tabashnik, B.E. (2020). Gene flow between Bt and non-Bt plants in a seed mixture increases dominance of resistance to pyramided Bt corn in Helicoverpa zea (Lepidoptera: Noctuidae). Journal of Economic Entomology, 113, 2041–2051.

Cerdeira, A.L., & Duke, S.O. (2006). The current status and environmental impacts of glyphosate-resistant crops. Journal of Environmental Quality, 35, 1633–1658.

Dederer, H.-G. (2020). Patentability of genome-edited plants: A convoluted debate. IIC International Review of Intellectual Property and Competition Law, 51, 681–684.

Franke, A.C., Breukers, M.L.H., Broer, W., Bunte, F., Dolstra, O., D'Engelbronner-Kolff, F.M., ... & Rutten, M.M. (2011). Sustainability of Current GM Crop Cultivation: Review of people, planet, profit effects of agricultural production of GM crops, based on the cases of soybean, maize, and cotton (Report 386). Plant Research International, Wageningen University and Research Centre.

Godt, C. (2018). Technology, Patents and Markets: The Implied Lessons of the EU Commission's Intervention in the Broccoli/Tomatoes Case of 2016 for Modern (Plant) Genome Editing. IIC International Review of Intellectual Property and Competition Law, 49, 512–535.

Haverkort, A.J., Boonekamp, P.M., Hutten, R., Jacobsen, E., Lotz, L.A.P., Kessel, G.J.T., ... & Visser, R.G.F. (2016). Durable late blight resistance in potato through dynamic varieties obtained by cisgenesis: Scientific and societal advances in the DuRPh project. Potato Research, 59, 35–66.

Kessel, G.J.T., Mullins, E., Evenhuis, A., Stellingwerf, J., Cortes, V.O., Phelan, S., ... & Van Der Voet, H. (2018). Development and validation of IPM strategies for the cultivation of cisgenically modified late blight resistant potato. European Journal of Agronomy, 96, 146–155.

Kruger, M., Van Rensburg, J.B.J., & Van den Berg, J. (2012). Transgenic Bt maize: Farmers' perceptions, refuge compliance, and reports of stem borer resistance in South Africa. Journal of Applied Entomology, 136, 38–50.

Lamichhane, J.R., Bischoff-Schaefer, M., Bluemel, S., Dachbrodt-Saaydeh, S., Dreux, L., Jansen, J.-P., ... & Malausa, T. (2017). Identifying obstacles and ranking common biological control research priorities for Europe to manage most economically important pests in arable, vegetable, and perennial crops. Pest Management Science, 73, 14–21.

Lotz, L.A.P., van de Wiel, C.C.M., & Smulders, M.J.M. (2018). How to assure that farmers apply new technology according to Good Agricultural Practice: Lessons from Dutch initiatives. Frontiers in Environmental Science, 6, 89.

Lotz, L.A.P., van de Wiel, C.C.M., & Smulders, M.J.M. (2020). Genetic engineering at the heart of agroecology. Outlook on Agriculture, 49, 21–28.

Lotz, L.A.P., van deWiel, C.C.M., & Smulders, M.J.M. (2014). Genetically modified crops and sustainable agriculture: A proposed way forward in the societal debate. NJAS Wageningen Journal of Life Sciences, 70–71, 95–98.

Lotz, L.A.P., Wevers, J.D.A., & van der Weide, R.Y. (1999). My view. Weed Science, 47, 479–480.

Mortensen, D.A., Egan, J.F., Maxwell, B.D., Ryan, M.R., & Smith, R.G. (2012). Navigating a Critical Juncture for Sustainable Weed Management. Bioscience, 62, 75–84.

NASEM (National Academies of Sciences, Engineering, and Medicine). (2016). Genetically Engineered Crops: Experiences and Prospects. The National Academies Press.

Parra, L., Maisonneuve, B., Lebeda, A., Schut, J., Christopoulou, M., Jeuken, M., ... & Michelmore, R. (2016). Rationalization of genes for resistance to Bremia lactucae in lettuce. Euphytica, 210, 309–326.

Roe, D., & Brokaw, R. (2007). Intellectual property rights applicable to fruit trees and the likely effects on regional and global avocado industries. In Proceedings of the VI World Avocado Congress (Actas VI Congreso Mundial del Aguacate), ISBN 978-956-17-0413-8.

Schaart, J.G., van de Wiel, C.C.M., Lotz, L.A.P., & Smulders, M.J.M. (2016). Opportunities for products of New Plant Breeding Techniques. Trends in Plant Science, 21, 438–449.

Smulders, M.J.M. (2020). Then and Now: A Scientific View on Plant Breeding and Technological Innovation. In Sustainable Agriculture: The Role of Plant Breeding Innovation (pp. 21–25). Institute on Science for Global Policy. ISBN 978-1-7334375-2-3.

Thornby, D., Werth, J., & Walker, S. (2013). Managing glyphosate resistance in Australian cotton farming: Modelling shows how to delay evolution and maintain long-term population control. Crop and Pasture Science, 64, 780.

Van de Wiel, C.C.M., Schaart, J.G., Lotz, L.A.P., & Smulders, M.J.M. (2017). New traits in crops produced by genome editing techniques based on deletions. Plant Biotechnology Reports, 11, 1–8.

Wesseler, J., & Purnhagen, K.P. (2020). EU Regulation of New Plant Breeding Technologies and their Possible Economic Implications for the EU and Beyond. Bayreuth Work. Paper Ser. Food Law.

Würtenberger, G., Van Der Kooij, P., Kiewiet, B., & Ekvad, M. (2015). European Union Plant Variety Protection (2nd ed.). Oxford University Press.

Index

www.ingramcontent.com/pod-product-compliance
Lightning Source LLC
LaVergne TN
LVHW021256210726
843527LV00003B/196